SIFTING
THE
TRASH

SIFTING THE TRASH

SIFTING THE TRASH

A HISTORY OF DESIGN CRITICISM

ALICE
TWEMLOW

The MIT Press
Cambridge, Massachusetts
London, England

This book was set in PF DIN Pro by The MIT Press. Printed and bound in Canada.

Library of Congress Cataloging-in-Publication Data
Names: Twemlow, Alice, author.
Title: Sifting the trash : a history of design criticism/Alice Twemlow.
Description: Cambridge, MA : The MIT Press, 2017. | Outgrowth of the author's thesis (doctoral--Royal College of Art, 2013) under the title: Purposes, poetics, and publics : the shifting dynamics of design criticism in the US and UK, 1955-2007. | Includes bibliographical references and index.
Identifiers: LCCN 2016034155 | ISBN 9780262035989 (hardcover : alk. paper)
Subjects: LCSH: Product design--Social aspects. | Design--Public opinion.
Classification: LCC TS171.4 .T89 2017 | DDC 658.5/752--dc23 LC record available at https://lccn.loc.gov/2016034155

10   9   8   7   6   5   4   3   2   1

CONTENTS

ACKNOWLEDGMENTS

This book began as a PhD dissertation through the Royal College of Art/Victoria & Albert History of Design course. It was an extreme privilege to work with my supervisors, Jeremy Aynsley and David Crowley, whose research, thinking, and teaching set the gold standard for our field. I am also grateful for thoughtful input by other RCA/V&A tutors and the students during our work-in-progress sessions.

While undertaking my doctoral studies, I was also the founding chair of the MFA in Design Criticism and the MA in Design Research, Writing & Criticism at the School of Visual Arts in New York. For this remarkable opportunity, my heartfelt thanks go to Steven Heller and David Rhodes at SVA. I am also indebted to my colleagues, students, and alumni, whose critical thinking about design deployed in all manner of modes, from the guerrilla bodega exhibition to the reality television series, continually inspires me.

This book benefits from the expertise of archivists and librarians who have facilitated archival research at the Archive of Art and Design, Bard Graduate Center library, University of Brighton Design Archives, Design Museum Archive, Getty Research Institute, Special Collections and University Archives, University of Illinois at Chicago, Institute of Contemporary Arts Archive, National Art Library School of Visual Arts library, and the University of the Arts library, among others.

I would like to express my gratitude to all the interviewees who so generously shared the insights and recollections that inform this research. They are as follows: Deborah Allen, Mary Banham, Stephen Bayley, Ralph Caplan, Claire Catterall, Sheila Levrant de Bretteville, Anthony Dunne, Simon Esterson, Richard Farson, Merrill Forde, Ken Garland, Richard Hamilton, Dick Hebdige, Mark Kingsley, Peter Murray, Eli Noyes, Rick Poynor, Fiona Raby, Deyan Sudjic, Jane Thompson, Judith Williamson, Jon Wozencroft, and Peter York.

A 2015 Production and Presentation Grant from the Graham Foundation for Advanced Studies in the Fine Arts enabled the images in this book to be sourced and licensed, and funded some color reproductions. Particular thanks go to Enya Moore, my valued research assistant, for doggedly tracking down the images, and to Sonia Mangiapane for taking some of the photographs. Among those individuals and institutions who helped me in gathering and gaining permission to publish images are: Oliver Allen, Architectural Press Archive/RIBA Collections, Mary Banham, Ralph Caplan, Nigel Coates, Design Council Archives, University of Brighton Design Archives, Design Museum, Design Curial, Dunne & Raby, F&W, Ken Garland, Graphic Thought Facility, Richard Hamilton Estate, Dick Hebdige, Alvin Lustig Archive, Mowat and Company, Peter Murray, Eli Noyes, Jane Thompson, and Armin Vit.

Two of my esteemed SVA colleagues, Adam Harrison Levy and Russell Flinchum, kindly read the manuscript in its various states of disarray, and offered helpful advice. I was also lucky enough to have Thomas Weaver, editor at the Architectural Association, and the design historian Pat Kirkham as my PhD external examiners; their insightful comments have contributed meaningfully to the final shape and complexion of this book. I would also like to thank Emily King and Rick Poynor for being my long-term, and long-distance, mentors.

Many thanks are due to Roger Conover at the MIT Press for believing in this book, and in particular to Victoria Hindley for patiently shepherding me through the process of its production. I am also indebted to my skillful editor Gillian Beaumont, to Margarita Encomienda for the sensitive design of the book, and to Matthew Abbate and everyone else at MIT involved in the complex project of bringing a book to press.

Parts of chapter 2 have been published as essays in the journal *Design & Culture*, volume 1, number 1 (Berg, March 2009), as "I Can't Talk to You if You Say That: An Ideological Collision at the International Design Conference at Aspen, 1970," and in the book *The Aspen Complex*, ed. Martin Beck (Sternberg Press, 2012), as "'A Guaranteed Communications Failure': Consensus Meets Conflict at the International Design Conference in Aspen, 1970." Special thanks to their editors, Elizabeth Guffey, Martin Beck, and Leah Whitman-Salkin, respectively.

Finally, I would like to thank my mother, Cayla Twemlow, and my father, Graham Twemlow, for their immeasurable love and support, and my wonderful husband David Womack for his guidance, humor, and patience, which have provided much-needed intellectual and emotional ballast over the many years that this project has cohabited with us. Very last of all, thanks to our son Otto Gray Womack, an informed critic of Lego, Minecraft, and playgrounds, among other aspects of the designed environment, for all the happy distraction he provides.

This book is dedicated to the design critic Deborah Allen (1924–2014). I am very grateful that I had the chance to interview the woman who coedited *Industrial Design* magazine according to the dual measures of poetry and pragmatism, reviewed 1950s cars with such panache, took care of five children, and wrote the line "hers is a lush situation." It really is. My thanks to all.

Amsterdam, 2016

INTRODUCTION

Debut and demise, purity and pollution

Design criticism operates at the very brink of the landfill site, salvaging some products from its depths, but also hastening the descent of others through its condemnation or indifference. Like the contractors and scavengers who amass, and comb through, Victorian London's rubbish heaps in *Our Mutual Friend*—hoping to find treasure in the "Coal-dust, vegetable-dust, bone-dust, crockery dust, rough dust and sifted dust,—all manner of Dust," so design curators and critics amass and comb through the looming detritus of contemporary society, temporarily arresting the progress of products on their journey from factory to junkyard, and diverting them toward a spotlit, white plinth or a glossy, double-page spread.[1]

Criticism, as the exercise of making distinctions between things, stems from a long tradition of liberal humanist criticism, encapsulated by literary critic I. A. Richards as "the endeavour to discriminate between experiences and to evaluate them" or, by R. P. Blackmur, as the endless search "with every fresh impulse or impression for better names and more orderly arrangements."[2] Tied etymologically to *discernment*, the act of distinguishing between things and separating by sifting, the term *discrimination* is fundamental to design criticism, which has sought to identify, sort, categorize, and assign values to things, to tidy up; but by doing so, it "conjures up disorder together with the vision of order, dirt together with the project of purity," as Zygmunt Bauman has observed.[3]

Like its close relations in the fields of art, architecture, and literature, product design criticism is driven by all manner of motivations and priorities, often operating simultaneously, which can be didactic, aesthetic, interpretative, promotional, oppositional, poetic, political, ideological, or cultural. What distinguishes product design as subject matter for criticism is its sheer ubiquity—it is estimated that the online retailer Amazon offers more than 480 million different products, for example—and the force of its impact on the environment and on society, through the ways in which these products are made, promoted, consumed, and disposed of, and also how they are regarded in the popular imagination.

Product design criticism is characterized by its concentration on the new, ever since mass production conferred value upon the novel commodity.[4] Product design criticism reinforces the fetishization of box-freshness by continuing to evaluate goods that are just off the production line conveyor belt or, worse, just off the press release, as framed and presented by their designers, manufacturers, retailers, and promoters. By doing so, it contributes to the velocity of the production-consumption-disposal cycle, and to the volume of the 1.3 billion tons of municipal solid waste produced globally per year. Waste that, in English literature professor Brian Thill's conception, "lays thick blankets of our chemical age across the entire planet, into every rocky outcropping, to the bottom of every sea's floor, nestling in the trees and bogs and pools of the world. It's in the air, in the water, in the yard sales brimming with kitsch, in houses stuffed to the rafters with rubbish, in outer space, spreading out in invisible clouds of toxic chemicals, and piling up in immense mountains of garbage stacked in trash-bricks below ground at Fresh Kills or Puente Hills or a thousand other dump sites."[5]

When design is considered from this perspective, as the waste it will become, then, as design theorist Ben Highmore has observed, "it is hard not to see global warming and climate change as a consequence of a variety of design processes, design values and design products."[6] French social scientist Bruno Latour summarizes the predicament most succinctly when he says, "Between modernizing and ecologizing, we have to choose."[7] Latour's demand on the collective critical conscience has become even more urgent in the light of the recent election of entrepreneur Donald Trump as the American president and his stated intention to withdraw the US from the Paris Climate Accord, an agreement made by 193 countries to control the increase in industrial emissions that are heating the planet, which became international law on November 4, 2016.

A gathering acknowledgment of the design industry's complicity in climate change is generating demand for consideration of other moments in a product's lifecycle, apart from its birth, such as how it might be used by someone over time, what happens after its period of usefulness is over, and what happens when it is disposed of. As Czech philosopher Vilem Flusser has noted, "waste is becoming more and more interesting."[8]

Increasingly, we are seeing the "new and shiny objects of our age" not only as the products of design ingenuity, manufacturing labor, and consumer desire but also, in terms of their afterlives, as the "rusting, splintered, discarded" husks they will become, and not only embedded in their present time and space but "carom[ing] off the edges of the present and into the past and future," as Thill puts it.[9]

Branches of archaeology, social studies, and the humanities such as garbology, discard studies, eco-criticism, geophilosophy, and sustainability studies all provide models for research into post-use manufactured goods and waste systems, in the context of the geological epoch of the Anthropocene. Geobiologists such as Reinhold Leinfelder stress the importance of discussing different conceptions of time and how they interact with each other, such as cosmic, evolutionary, cultural, technological, societal, and individual time scales.[10] Ecological theorist Timothy Morton proposes the term "hyperobject" to help us think of things in terms of their massive distribution in time and space, relative to humans. For example, "a hyperobject could be the very long-lasting product of direct human manufacture, such as Styrofoam or plastic bags, or the sum of all the whirring machinery of capitalism."[11]

While Morton's "hyperobjects" involve profoundly different temporalities than the human-scale ones design historians and critics are used to working with, as recent conferences, research centers, and academic courses of study attest, the field of design studies is beginning to attend to an expanded notion of time scale in relation to design, including product lifecycle analysis and what in architecture is termed "post-occupancy evaluation."[12] A closer examination of what happens when a designed product becomes trash, of the social behaviors, politics, infrastructures, mechanisms, and economies that shape and gather around refuse and its disposal, could surely enrich our understanding of design culture and provide a much-needed critique of currently dominant labels that mislead with their deflection of attention away from the physicality of waste, such as the supposed immateriality of information, the "cloud," service design, "innovation culture," and the "creative economy."

It is a timely moment, therefore, for a reexamination of the history of product design criticism through the metaphors and actualities of the product as imminent junk and the consumer as junkie. Through its excavation of instances when critics attempted to question design's impact on the physical environment and the social psyche, and its examination of experimental modes of practice, this book offers an array of precedents for how an evolving practice of design criticism might be conducted in the future.

Sifting through the trash and diagnosing sickness as metaphors

Design criticism performs an evaluative role in which it allots products positions on a value hierarchy that extends from the reviled status of trash at one end to the revered status of treasure at the other. These positions are only ever temporary, however; as Susan Strasser has noted, trash is a dynamic category, the status of goods is in constant flux.[13] They are dislodged and relodged by the combined and shifting forces of the market, the design canon, and taste. British design and architecture critic Reyner Banham observed that "all transient consumables slide slowly down the parallel scales of social esteem and actual cash value until they bottom out as absolute rubbish. At that point, however, they are not necessarily discarded, but may suddenly leap to the top of both scales."[14]

Design critics have taken different standpoints with regard to this sifting of the trash and assigning of value. In the mid-1950s through the 1960s, Banham sought to develop what he termed "an aesthetics of expendability," as a way to celebrate disposable consumer goods and to counter an entrenched and elitist value system based on durability and permanence. He rescued from the trash pile things like potato crisp packets, transistor radios, bank notes, sunglasses, Californian surfboards, paperback book covers, and the decoration of ice cream trucks. Banham identified his subject matter as the kinds of new, cheap, mass-produced products that figured in people's lives—things designed not to "be treasured, but to be thrown away." In the absence of the right language or "the intellectual attitudes" for comprehending and analyzing what he called "living in a throwaway economy," Banham attempted to formulate them himself through the use of new diction, metaphors, syntax, as well as a self-reflective use of new methods necessary for participating in what he called "the extraordinary adventure of mass-production."[15]

By the late 1960s, the extraordinary adventure of mass production had proved disillusioning for other design critics. The 1970 International Design Conference at Aspen was the occasion for a protest by students and environmental activists in which they called attention to the hypocrisy of the liberal design establishment and its superficial interest in the environment. Garbage was used throughout the conference both as a metaphor and as a physical prop to support a sentiment encapsulated by commune dweller and Ecology Action's founder Clifford Humphrey: "If an item is made to be wasted, to be dumped on a dump, then don't make it!"[16]

In the 1980s, British design curator Stephen Bayley sent the design objects he thought "disgusting" back to landfill, by placing them on upturned dustbins in his 1983 Boilerhouse Project exhibition on "Taste." And in the 1998 ICA "Stealing Beauty" exhibition, the curator Claire Catterall revisited the trashcan by selecting for display examples of contemporary design in 1990s London made from, and inspired by, the detritus of everyday life. Most of the exhibits reanimated found materials and rubbish. Catterall wrote in the exhibition catalog: "If you look at the work it is, essentially, just a load of old tat. Supermarket trolleys, lottery tickets, flyposters, blue and white table china, office signage, 2 x 4 ply, football terrace chants, council estate maps, the work is littered with things stolen from the landscape of our everyday lives."[17] And finally, design critic and editor Rick Poynor consigned the verbal junk of early-twenty-first-century design blogging back into the computer's trash icon.

Running alongside, and sometimes in opposition to, this imperative to sift the trash and assign value, throughout the latter half of the twentieth century, was the compulsion on the part of design critics to detect, diagnose, and even provide therapy for the sickness of a consumer society perceived to be binge-consuming itself into a state of collective psychosis.

The postwar British design establishment wielded "disinfectants and anaesthetics" against "socialized welfare state man" in their efforts to cleanse him of poor taste.[18] When Jean Baudrillard wrote a paper for the International Design Conference at Aspen in 1970, he critiqued the American design community's alleged concern about the environment as "naive euphoria in a hygienic nature," and the establishment's focus on environmental pollution as a means of seeking to protect itself from the equally polluting influences of communism, immigration, and disorder. Dick Hebdige, Judith Williamson, and other critics in 1980s Britain repeatedly referenced sickness, and particularly mental illness, to characterize the effects of design, and specifically style, on society. They critiqued 1980s British design for its collusion with Thatcherite enterprise culture and its provision of "institutionalized therapy" in the form of more consumer goods for the very consumerist sickness it had helped engender. In the late 1990s, the proponents of critical design Dunne & Raby created "placebo objects" to draw attention to, and provide a salve for anxieties about, the electromagnetic fields caused by electronic products in people's homes.

Design criticism's relationship to its publics, imagined and real

Both the winnowing and the diagnostic roles of criticism imagined a public that needed its values deciding and its sickness identified. In their tone of address and themes covered, the design critics considered in this book conceived of their public as a body of citizens that might need educating in relation to, castigation for their complicity with, or protecting from, the design industry and its products. Throughout the latter half of the twentieth century one can trace three main conceptions of the public addressed by design critics: a public that needed to be educated on behalf of the larger design enterprise in order to make, sell, and buy better products; a public that needed to be protected from the machinations of commerce and advocated for; and a public that, through their reading and self-publishing behaviors, was seen as both a fulfillment of, and a threat to, the goals of design criticism itself.

The first—the didactic critical imperative—had its roots in deep-set traditions of design reform dating back to the mid-nineteenth century and to the notion of public service in Britain and public good in the US, which shaped much thinking about design in the first half of the twentieth century, but reached fruition in the interwar and immediate postwar periods. The second—the protective imperative—gained traction with the development of consumer protection organizations and publications such as *Consumer Reports* and *Which?* in the 1950s. The third category, an amateur, or "do-it-yourself," mode of design criticism, became most pronounced with the advent of blogs in the early 2000s and the opportunity they afforded for members of the public to launch their own publications and contribute comments to others, posing a dilemma for the kind of design criticism which had wanted to empower its public to perform critique, but whose own power, authority, and gatekeeping role was increasingly eroded in the process.

While these conceptions of the public can be historically located, in fact all three coexisted and exchanged predominance throughout the entire period. The style and life-style discourse of British 1980s design publications can be seen as a continuation of the instructional work of the 1950s design establishment who wanted to teach people how to acquire taste and how to live, and continues to this day in many exhibitions and blogs. The democratizing impulse to share the strategies and insights of the critical apparatus, while most apparent in the recent era of blogging, was also a concern of the editors of *Industrial Design* magazine in the 1950s, who sometimes published articles sent in by readers, and saw their role as enabling their readers to perform their own criticism.

Jürgen Habermas, the German philosopher who conceptualized public life in the interwar period, has described a "public sphere," where people "behave neither like business or professional people transacting private affairs, nor like members of a constitutional order subject to the legal constraints of a state bureaucracy," but, rather, as "citizens."[19]

Habermas's conception of an engaged citizenry (differentiated from consumers) and an arena assembled for the purposes of debate and for forming public opinion (separate from the commercial transactions of the marketplace) has provided much of the impetus for the performance of design criticism. Through the adoption of a particular voice, argument, and attitude in the artificially staged environment of a magazine or blog, or more literally at the podium of a lecture hall (where many pieces of written criticism begin), a design critic performs in imagined dialogue with former critics, peers, and the designers whose work she discusses, and before the imagined audience of the reading public.

In addition to the imagined presence of a larger public that may or may not have come directly into contact with the ideas expressed in the criticism, there was an actual community of individuals who subscribed to, bought, visited, attended, or otherwise sought out the vehicle through which the criticism was disseminated. Even though the task of learning what everyday goods mean to the people who consume them is, according to Jeffrey Meikle, one of design history's "central challenges," in the case of the history of design criticism it is possible to access traces of engagement with the products of criticism, through letters to the editor and comments in the case of a publication or blog, and questions or protests in the case of a conference.[20] As the author Deanya Lattimore has observed of the publishing industry, "There is no pre-existing public. The public is created through deliberate, wilful acts: the circulation of texts, discussions and gatherings in physical space, and the maintenance of a related digital commons. These construct a common space of conversation, a public space, which beckons a public into being."[21]

The materiality of criticism

In its original context, design criticism borrows energy from everything around it—in the case of written criticism, the work of other writers in a magazine, newspaper, or blog, the juxtapositions between articles and advertising, the choice of images, the editorial framing, the letters to the editor, the pacing of the sequencing, and the layout of the pages. A piece of text exists in space, is a designed entity made of materials, and is subject to similar economic pressures as other designed products. By returning to examine an article in its original location in a publication, one can piece together the live community in which it had a particular purpose and intention, in which it mattered. As a social group, what did the publishers, editors, writers, photographers, art directors, advertisers, and readers of a publication care about at that time, and why?

Critical documents can be seen as nodes in larger networks comprised of writing, designed objects, ideas, and people. The historian of criticism seeks to reconnect the links between these entities and to reimagine the social geographies that gave rise to their creation. Rather than trying to create some perfect reconstruction of an article in situ, like a period interior, however, the historian must acknowledge the fluidity and instability of the intertextual framework, and the ways in which looking at historical examples of criticism necessarily involves a negotiation with the ways they have been previously interpreted and the concerns of the present. An intertextual reading is thus invoked here in two ways: first in the sense that Julia Kristeva coined the term, as a means of appreciating a text as an author's production of a mosaic of references to, quotations from, and implicit dialogues with other texts.[22] The term "intertextuality" can also, secondly, be used to describe the way in which a reader of a text draws on his experience with other texts, and makes connections between these various texts and the present text being experienced.[23]

Ann Sobiech Munson's analysis of Lewis Mumford's review of the Lever House building published in the *New Yorker* on August 9, 1952, is a good example of a historian's attempt to reconstruct the social life of a piece of writing through a detailed description of its location in the magazine, abutted by advertisements for synthetic fabrics, air travel, and the 1952 Lincoln (with its "3,721 square inches of glass"), other articles, reviews, cartoons, and pieces of fiction, and how it would have been received by "a recognizable *New Yorker* reader, habitat, and geography of the mind." Sobiech Munson also compares the Lever House article to others written by Mumford in his ongoing column "The Skyline," and to other reviews of the building published in contemporaneous magazines. In doing so, she makes a case for the vital role of Mumford's writing in the mid-century American understanding of modernism: not merely "representing" architecture but actively participating in the way it was understood and adding to the construction of the architectural subjects, thus contributing to the social-historical record of the built artifact. "Writing becomes the object of study," one that not only reflects the built object Lever House but also inflects back into the icon Lever House and becomes complicit in the construction of the world it inhabits.[24]

Design historians and theorists often use journalistic writing as raw material for the writing of history and theory.[25] Repositioning writing about design from its marginal location, as source material, to a more central location where it becomes the object of study brings with it specific methodological challenges. In 1984, design historian Clive Dilnot suggested that "a history of the rise of the design journal as the vehicle for projecting the ideology or value of 'design' would be an enormous contribution to understanding the profession's self-promotion of design values. To map the changing values, idea, and beliefs expressed or communicated in text and graphic layout could, in a sense, map the history of the professions. Is the history literally contained in the glossy pages of *Domus* or *Industrial Design*?"[26]

While there certainly is much to be contributed to design history by gleaning information from "the glossy pages of *Domus* or *Industrial Design*," Dilnot's view of the design journal as a "projector" of the values of the design profession is limiting. Magazines encompass the contrary views of, and complex relationships between, publishers, editors, writers, readers, and subjects, in ways far more heterodox than Dilnot's assessment suggests. Product design critics of the early 1950s attempted to balance the perceived needs of their various constituencies—designers, manufacturers, policymakers, and consumers—with the aims of the commissioning magazine and their personal literary ambitions. A more nuanced study of design magazines should account for the political, social, and economic pressures that shape them, the variety of voices and opinions expressed, and especially for the moments of resistance to design's ideologies and values that occur on their pages.

This flux of contradictory ideas, imperatives, and interpretations might be termed the "dynamics of criticism." As M. C. Lemon has observed, history involves the examination of "the genuine interplay (rather than meaningless juxtapositions) of individuals with each other and with a multiplicity of phenomena such as groups, parties, institutions and ideas."[27] Cultural studies, too, identifies such interplay. In his inaugural address as the first director of the Centre for Contemporary Cultural Studies at the University of Birmingham in 1963, the sociologist Richard Hoggart outlined an approach to what he provisionally called "Literature and Contemporary Cultural Studies." Hoggart identified four main foci for this nascent discipline: writers and artists (how do they become what they are, and what are their financial rewards?); audiences (what expectations do they have, and what background knowledge do they bring?); opinion-formers, guardians, the elite, the clerisy (where do they come from, and what are their channels of influence?); and the organizations for the production and distribution of the written and spoken word (what are their natures, financial and otherwise?). Lastly, Hoggart spoke of an urgent need to find out more about what happens when all four shaping forces interact, "about interrelations between writers and their audiences, and about their shared assumptions; about interrelations between writers and organs of opinion, between writers, politics, power, class and cash."[28] Hoggart's summary of the concerns of the CCCS aptly describes the project of a historian of design criticism, and such questions have impelled the research presented in this book.

The poetic, literary, and narrative qualities of design criticism

A major component in the dynamics of criticism is a quality which all of the critics studied in this research displayed, and that is their acute sense of their own writerly abilities, their delight in the nuances and texture of language, and their sympathy for, and ambition to achieve, literary status. Literary critic Terry Eagleton has written that "few words are more offensive to literary ears than 'use,' evoking as it does paperclips and hair-dryers. The Romantic opposition to the utilitarian ideology of capitalism has made 'use' an unusable word: for the aesthetes, the glory of art is its utter uselessness."[29] And when the *Los Angeles Times* automotive writer Dan Neil won the 2004 Pulitzer Prize for criticism, a *New York* magazine theater critic responded: "If you write about cars it is reportage. It is not criticism. Cars are utilitarian things. You might as well be a critic of kitchen utensils."[30]

Such prejudices represent some of the anxiety and tension that by turns stultified and fueled design criticism in the latter half of the twentieth century. For, of course, paperclips, hair-dryers, cars, and kitchen utensils are design criticism's subject matter—its stock-in-trade—and despite their personal romanticism, idealism, literary ambitions, or aesthetic, social, and moral imperatives, design critics have been acutely conscious of a perceived pressure to engage with how such things are used.

It could be said that the glory of design is its utter use*ful*ness. It is mostly experienced in everyday conditions rather than in those spaces separated for enabling transcendent thought, like theaters or art galleries. So design criticism, unlike most other forms of criticism, is often characterized by its focus on the ordinary and the ephemeral, although, as art critic Dave Hickey reminds us, seeking to overcome the incommensurability of sensory experience, let alone the "enigmatic whoosh of ordinary experience," is no easy task. In his own attempts, Hickey found himself "slamming ... against the fact that writing, even the best writing, invariably suppresses and displaces the greater and more intimate part of any experience that it seeks to express."[31]

The fact that design is so centrally located within arenas of economic exchange fundamentally affects what is written about it. Design criticism is unusual, possibly unique, among other genres of criticism in its proximity to, and need to evaluate, the processes of manufacture, retail, and distribution. Other genres of criticism are subject to the same economic realities, and operate in the same, or at least overlapping, commercial spheres. Literary, film, and art criticism (and not just reviews) all have an impact on sales of their respective products, but the commercial implications of such influences are rarely discussed. Although the depth and quality of economic discussion in design criticism are usually slight, design critics do have to consider the mechanics of making and selling, in more explicit terms than critics in other genres.

This does not mean that design critics are comfortable with doing so. Between 1954 and 1956, the French professor of literature Roland Barthes wrote a monthly column about modern products for the Paris literary magazine *Les Lettres Nouvelles*. His article on the 1955 Citroën exposes his ambivalence toward the grubbiness of retail culture. Barthes discussed the display and marketing of the car, but his repulsion is evident in his condemnation of the speed of the process of its mediatization—a process which he saw as wholly symbolic of petit bourgeois values. Yet he attended the car show; reporting from the scene of the Déesse's commercial exhibition and on the details of its mediation was necessary to a full discussion of its symbolic value. By contrast, it is not necessary for literary critics to report on, say, the circumstances of a book's display at the Frankfurt Book Fair.

Deborah Allen, automobile reviewer for *Industrial Design* in the 1950s, did not hide her dislike of the machinations of the Detroit auto industry, yet she completed her formal analysis of the latest car models with discussion of the economic strategies of their manufacturers, sales figures, and the ways they were marketed. In his scrutiny of products such as the *Habitat* catalog or *Face* magazine in the 1980s, critic and theorist Dick Hebdige combined his semiotic readings of these products' imagery with appraisals of the way they shaped their readers' behavior in the marketplace.

Design criticism is often torn between its need to report from the bustle of the bazaar and its cultural ambition to contribute to a discourse that rises above the arena of business transaction, dealing with seemingly loftier themes of inspiration, emotion, morals, and human values and in the evocation of what architecture critic Reinhold Martin has termed "other, possible worlds."[32]

Design criticism is also tied to the design industry and the marketplace through the means by which it is generated, funded, and broadcast. There are few instances of truly independent design criticism—perhaps a solely publicly funded institution such as the BBC, or a section of a newspaper that contains no design-related advertising, or a self-coded blog or website, might count as such. Most design criticism during the period under consideration in this book, however, was commissioned, paid for, and distributed by companies, institutions, and nonprofit organizations and grants that were supported and sponsored by commercial design enterprise. Thus the impossibility of design criticism's true disinterestedness sits uneasily with the idealism of its noncommercial, anticapitalist motivations, such as literary ambition, the desire to oppose and resist, and the search for social and political justice.

Many believe that design criticism is so deeply entrenched within the design industry, so closely tied to its professional goals, that its ultimate effect will always be promotional rather than critical. Social theorist Andrew Wernick has portrayed an emerging culture in the 1960s and 1970s whose communicative processes were gradually being saturated in the medium of promotion. He argued that neither satire nor critique is immune from the process it may seek to destroy through laughter or pointed insight: "Once we are communicating at all, and especially in public, and therefore in a medium which is promotional through and through, there is no going outside promotional discourse. These very words are continuous with what they are seeking to distance themselves from. To paraphrase what Derrida remarked of textuality in general: there is no hors-promotion."[33]

The critics discussed in this book attempted to find an escape from the "no hors-promotion" conundrum through their use of rich metaphorical and poetic language, as a means both to transcend the banal functionality of the products they dealt with and to get even closer to them. When Roland Barthes and Deborah Allen wrote about cars, they summoned ethereal and religious imagery. Reviewing a 1955 Buick, Allen suggested that the beholders should suspend their disbelief as they would upon encountering solid wooden clouds on the underside of a canopy of state in baroque cathedral architecture, and "accept the romantic notion that materials have no more weight than the designer chooses to give them." Barthes also used cathedral architecture as point of comparison for his analysis of the D.S. 19: "I think that cars today are almost the exact equivalent of the great Gothic cathedrals: I mean the supreme creation of an era, conceived with passion by unknown artists, and consumed in image if not in usage by a whole population which appropriates them as a purely magical object."[34] Then, like Allen, Barthes took the reader on a sensory exploration of the car's surfaces, which he had observed being enacted by consumers at car shows: "yet it is the dove-tailing of its sections which interest the public most: one keenly fingers the edges of the windows, one feels along the wide rubber grooves which link the back window to its metal surround. There are in the D.S. the beginnings of a new phenomenology of assembling, as if one progressed from a world where elements are welded to a world where they are juxtaposed and hold together by sole virtue of their wondrous shape."[35]

Barthes wrote about the car's interior from the perspective of the driver, and compared the levers to "utensils" and the dashboard to a homely kitchen environment: "The dashboard looks more like the working surface of a modern kitchen than the control room of a factory; the slim panes of matt fluted metal, the small levers topped by a white ball, the very simple dials, the very discreetness of the nickel-work, all this signifies a kind of control exercised over motion rather than performance. One is obviously turning from an alchemy of speed to a relish in driving."[36] Magical and spiritual allusions such as "alchemy," Gothic cathedrals, and "wonder" abound in this article, and diaphanous evocations such as "dissipation," motion, and airiness exist in tense juxtaposition with more technical, ergonomic, substantial, and humdrum points of reference such as "heavy bumper," "sturdy post," "utensils," "kitchen," and "wide rubber grooves."

Such tensions reflect the fraught nature of design critics' predicament as writers with the potential to create poetry, but also as commentators with a responsibility to explain, evaluate, and even sell. When Rick Poynor took the Speak Up bloggers to task in 2007, one of his main bones of contention was their lack of sensitivity to language. He listed "quality of writing style" as one of eight key tenets of good criticism, in contrast to an admission by Mark Kingsley that the Speak Up blog contained a lot of "shitty prose." In the instances explored in this research where language was abandoned in favor of atmospheric impressionism, in the case of the "powerhouse::uk" and the "Stealing Beauty" exhibitions in the late 1990s, and the agency of mute objects in the case of Dunne & Raby's criticism-embodying products of the early 2000s, their publics were confused. Critic Judith Williamson, who reviewed the "powerhouse::uk" exhibition, took issue with its language—"babble," "blab," "meaningless chatter," and "self-congratulatory streams of dislocated words and circular messages," as she variously referred to it.[37] Michael Horsham, writing of "Stealing Beauty," noted that "this show is about our collective confusion and it follows that the things in it also, intentionally or unintentionally, concern that confusion."[38] When they made furniture and appliance hybrids, whose primary purpose was to question social and political values, the designer firm Dunne & Raby found that they needed to insert them into narratives in order to make their critiques legible. They created elaborate videos and publications with staged photographs, and even though the objects were meant to embody questions, ultimately they had to present their users with written questionnaires to elicit responses to such questions.

A note on the scope of this book, and chapter outline

The history of design criticism in the latter half of the twentieth century in the US and the UK is punctuated with self-reflective interruptions during which design critics were acutely self-conscious about their purpose, role in society, relationship to their publics, and use of critical techniques and formats. Each instance of interruption can be seen to spotlight a type of criticism that was new or coalescent in its time period, and articulated in implicit or explicit response to the perceived antagonism of the dominant concerns and values of design criticism as an established practice. In identifying five moments of historical discontinuity in the practice of design criticism, therefore, this book presents a kaleidoscopically reassembling, time-lapse portrait of the intellectual, stylistic, and material constitution of design criticism between the early 1950s and the early 2000s.

The US and the UK, and specifically New York and London, were among the centers of design practice, commentary, and publishing throughout the period under discussion. Choosing to focus on these locations allows for an examination of the flows and interruptions, the translations and misinterpretations, of ideas and influences between these two countries in a shared language. *Industrial Design* magazine in the US and *Design* magazine in the UK kept a sharp eye on one another's activities and the output of local design practice through their correspondents. They sometimes commissioned articles from each other's stable of writers, and often republished articles from each other's magazines. The fact that during this period of postwar reconstruction many British social and cultural critics were absorbed by American economic and cultural values also plays a part in the geographical delimitation of this book. Transatlantic interchanges were a feature of the International Design Conference at Aspen, which British critics visited as speakers, attendees, or reporters. In the 1970 edition of the conference, this two-way dialogue expanded to include an incongruous clash of cultures between representatives of the American liberal design establishment, mainly from New York, Californian environmental activists, and French left-wing philosophers, among which hostile constituencies the British design critic Reyner Banham attempted to mediate. The thread of US-UK exchange continues in my discussion of blogs in the early 2000s, when British critic Rick Poynor, writing in an American magazine, angered the members of a mostly American online design community with his dismissal of their contribution to criticism, although such geographical identities dissolve somewhat in the virtual space of an online forum.

This book begins in the immediate postwar era, when two major magazines dedicated to industrial and product design in its own right were founded (*Design*, in London, founded in 1949, and *Industrial Design*, in New York, launched in 1954), and the debate surrounding the purpose of design criticism became more evident and self-reflective. In the process of working out their own critical stances, writers, editors, and readers of such magazines raised questions about design criticism's utility in relation to design practice, social good, intellectual culture, political interests, the environment, and consumer protection and empowerment. Not all such questions were new to the period; they tapped into wider and sometimes centuries-old philosophical discourses on the role of critique in society, ranging from liberal humanist discussions of aesthetics and rhetoric to theoretical discussion of the pervasiveness of politics, the constitutive nature of language, and the contingency of meaning. The application of such discourses to design as subject matter was not entirely new to the period either. In the US, early-twentieth-century pragmatists such as John Cotton Dana, through his work as director of the Newark Museum, had embraced design as subject matter. In Britain, the social criticism of design manufacture by nineteenth-century design reformists such as John Ruskin and William Morris, and a plethora of design commentators in the early twentieth century, represents a kind of proto-design criticism. What was particular to the early 1950s, therefore, was the intensification of interest in industrial design as a topic, and the establishment of magazines devoted exclusively to industrial and product design.

The industrial design profession, which had been developed in the 1930s and 1940s, began, once postwar recovery was under way, to be both promoted and scrutinized more energetically, and it is the charged nature of the discussions that emerged during this period that makes this a viable starting point for this book. This was a time in which the forces of mass production, planned obsolescence, a maturing industrial design profession, governmental and institutional support for design, and unprecedented consumer demand for products all converged to extrude the processes, systems, and products of design into a recognizable shape, and into unavoidable subject matter for critical attention. The early 1950s is also an interesting launch point, since the journalistic impulses of certain editors at *Design*, and the very existence of an independent trade publication like *Industrial Design*, signaled ways in which design criticism might escape the institutional purview of the Council of Industrial Design in Britain and the Museum of Modern Art in the US. In the early 1950s, the genre of product design criticism began to define itself more distinctly and self-reflectively in a burgeoning array of printed media, events, broadcasts, and exhibitions, which marshaled new critics, editors, organizers, and publics.

Ending this study in the early 2000s allows for the inclusion of the arrival of online publishing and a consideration of its turbulent effects on the way design criticism has been conducted and consumed, as well as criticism's reconfigured relationship to democracy, authority, and professional status in the early years of the twenty-first century. Design criticism became increasingly fragmented and distributed across web media, with multiple micro-constituencies, rather than recognized publishers or institutions, initiating, hosting, and feeding the many simultaneous and rhizomatic conversations.

German philosopher Walter Benjamin has suggested how the metaphor of a "constellation" is better suited to a consideration of historical associations than a straight line representing an uncritical notion of progress across time. Benjamin's constellation links past events among themselves, and can link "what has been with the now"; its formation stimulates a flash of recognition in the anachronistic confluence between different time periods.[39] He believed that "[The historian who starts from this] records the constellation in which his own epoch comes into contact with that of an earlier one. He thereby establishes a concept of the present as that of the here-and-now, in which splinters of messianic time are shot through."[40] In the early twenty-first century, design criticism underwent emphatic and constitutive change. And yet, among its characteristics, such as its shape-shifting dispersal among media, discourses, and disciplines, one can discern "splinters" of earlier periods in its conception, as well as threads of continuity throughout the entire period under investigation. As this book will demonstrate, design criticism has always been a fugitive enterprise, inhabiting the interstices between recognized subject silos such as art, architecture, and social sciences, and, beyond publishing, exerting influences on the approaches, activity, and output of museums, institutions, professional associations, schools, publishing, research, and retail.

Lemon, who has proposed a framework for the study and writing of the history of political thought that reemphasizes the explicatory use of narrative, uses the term "event" to mean "a sequence of occurrences singled out for notice."[41] Lemon's examples of events include arguments, holidays, parties, elections, revolutions, evenings out, and journeys, all of which are to a large extent "deliberately planned orderings of occurrences."[42] One of the implications of the analytic principle of events is the necessity to "select out" events of import and "narrow down" their parameters, in order to "locate contexts of occurrences where meaningful sequences (that is, genuinely related temporalities), are to be found."[43] This book identifies significant events that represent moments of rupture in the history of design criticism, and then, within them, unfolds narratives that reveal the nature of tension, conflict, change, and evident self-consciousness about criticism.

Italian philosopher Giorgio Agamben reads Nietzsche's *Untimely Meditations* as dealing with the way in which true contemporariness is about "disconnection and out-of-jointness" with respect to the present.[44] Those who neither perfectly coincide with their time nor adjust themselves to its demands are "precisely through this disconnection … more capable than others of perceiving and grasping their own time."[45] In refusing the "demands" of the prevailing strains of design commentary in each of their periods of practice, all the critics discussed in this book used their "out-of-joint" perspectives on design to grasp their contemporary moment as fully as possible. Each of the "moments" of charged discussion about design criticism presented exemplifies a different critical voice, subject matter, technique, medium, and type of public engagement. The main concern, however, has been to pick the examples that best demonstrate moments of transition and change at which critics were most self-aware of both the means and the purpose of their criticism.

The chapter focuses of this book are as follows: a selection of articles published in the design magazines of the mid to late 1950s and early 1960s which forcibly activated a new set of values with which to engage with expendable, mass-produced product design; a protest at the International Design Conference at Aspen in 1970 which posed a challenge to the established conference lecture format and to a lack of political engagement on the part of the liberal design establishment; a set of articles by cultural critics that critiqued the prevailing celebratory commentary on style and lifestyle in 1980s London; an independent exhibition that offered an alternative view of contemporary design in contrast to government-endorsed design exhibitions in 1990s London, with an additional focus on an intensification of thought about the designed object as a potentially viable critical format; and, lastly, a debate between the authors of a US design blog and an established British design critic writing in *Print* magazine that highlights a rift between the energetic amateur impulses of blogging culture and the editorial values of traditional print media.

This book uses a broad definition of design criticism as a self-aware and subjective practice of interpreting, discerning among, encouraging, or resisting the various aesthetic, moral, environmental, or social repercussions of the ideas, activities, and outputs of the design industry. This definition is also based on the premise that design criticism is not confined to its written manifestations, and can also be conveyed in multiple media such as the event, the lecture, the exhibition, and the designed object itself. The kinds of criticism conducted through such activities as editing, oration, and debate, performance, the assembling and juxtaposing of objects, and the design process have different registers, textures, methods, and audience responses. Analysis of such modes, means, and sites of engagement contributes to a fuller understanding of criticism as a pervasive force exerting often invisible and unrecognized pressures on the ways in which design is developed, circulated and used.

For Latour, the critic's role is that of an instigator of public conversation: "The critic is not the one who debunks but the one who assembles. The critic is not the one who lifts the rugs from under the feet of the naïve believers, but the one who offers the participants arenas in which to gather."[46] His "Making Things Public: Atmospheres of Democracy" project, an exhibition and anthology of texts, considered the atmospheric conditions—technologies, interfaces, platforms, networks, and mediations—and the spaces in which communal public debate about things takes place:

> Scientific laboratories, technical institutions, marketplaces, churches and temples, financial trading rooms, Internet forums, ecological disputes—without forgetting the very shape of the museum inside which we gather all those membra disjecta— are just some of the forums and agoras in which we speak, vote, decide, are decided upon, prove, are being convinced. Each has its own architecture, its own technology of speech, its complex set of procedures, its definition of freedom and domination, its ways of bringing together those who are concerned—and even more important, those who are not concerned—and what concerns them, its expedient way to obtain closure and come to a decision.[47]

Since the occasions of critical debate examined in this book involve educators, philosophers, journalists, editors, designers, curators, conference organizers, artists, and activists, who deploy theory, reporting, lived experience, and ideology in combination, this broad view of design criticism's platforms helps to complicate an oft-invoked binary opposition between the so-called "academic" and "journalistic" variants of design criticism.

The apparent mutual distrust between these two cultures (academe and journalism) still underlies much discussion of criticism, typified by the terms and language used in a recent debate about the public accessibility of academic research, initiated by Rick Poynor ("The Closed Shop of Design Academia") and extended by Matt Soar ("Rick Poynor on 'Design Academics': Having His Cake and Eating It Too") and Peter Hall ("Changes in Design Criticism"), among others.[48] Peter Hall, writing from the perspective of a seasoned design journalist and scholar, reviews the similarities between the two fields and offers suggestions for how to move beyond this alleged divide, but by continued reference to books and magazines alone, it is hard to escape the "ivory tower" versus popular "marketplace of ideas" dichotomy. Extending the discussion beyond the restricted terms of a twentieth-century publishing paradigm allows for a more expansive conception of the evolution of design criticism in all the unexpected and unfamiliar forms it may inhabit, the concerns it may animate, and the publics it may speak for and with.

BUICK, though it is a year old, is the most revolutionary car on this spread. It is logical, but only by its own standards. It was not designed to sit on the ground or even roll on the ground; it is perpetually floating on currents that are conveniently built into the design. This attempt to achieve buoyancy with masses of metal is bound to have the same awkward effect as the solid wooden clouds of a Baroque baldachino; unless you like to wince a purist's wince at every Buick or baldachino, the best recourse is to accept the romantic notion that materials have no more weight than the designer chooses to give them. (Admittedly, this is hard when you witness the effect of a bump in the road on the Buick's heavy rear cantilever, which pretends to be diaphanous.) The Buick's designers put the greatest weight over the front wheels, where the engine is, which is natural enough. The heavy bumper helps to pull the weight forward; the dip in the body and the chrome spear express how the thrust of the front wheels is dissipated in turbulence toward the rear. Just behind the strong shoulder of the car, a sturdy post lifts up the roof, which trails off like a banner in the air. The driver sits in the dead calm at the center of all this motion; hers is a lush situation.

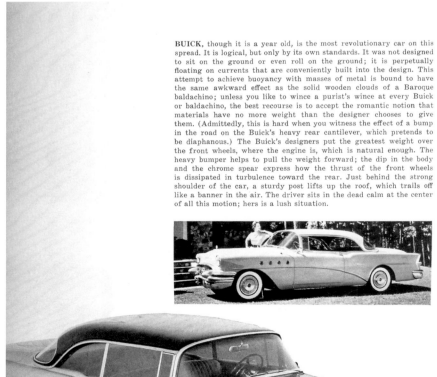

1

"A Throw-Away Esthetic":
New Measures and Metaphors in
Product Design Criticism,
1955–1961

One evening in early 1955, somewhere on US Route 1, while riding shotgun in a Buick Century, Deborah Allen had an epiphany. The auto critic and coeditor of *Industrial Design* magazine had been visiting a friend in Westport; as he drove her back to New York along the coastal road, and the setting sun reddened the Buick's chrome detailing, Allen suddenly saw how her friend "lived in his car and how he enjoyed it." As a public transport user and a skeptic of the exaggeratedly low-slung and streamlined cars of the period, she remembered being "amazed that there could be some sense in this car. It was a revelation."[1] Back in the magazine's Midtown office, Allen quickly typed up a report on her Olivetti Lettera 22. All the exhilarating impulsion of her recent ride is captured in a review that, unlike many of her others, seems to epitomize the era's most optimistic view of cars and what they promised in terms of mobility, modernity, and social progress. The Buick, she told her readers, "was not designed to sit on the ground or even roll on the ground; it is perpetually floating on currents that are conveniently built into the design." Allen's analysis of the way in which the car's styling reinforces its dynamics combines technical specificity with such an expressive and lyrical use of language that it verges on poetry: "The Buick's designers put the greatest weight over the wheels, where the engine is, which is natural enough. The heavy bumper helps to pull the weight forward; the dip in the body and the chrome spear express how the thrust of the front wheels is dissipated in turbulence toward the rear. Just behind the strong shoulder of the car, a sturdy post lifts up the roof, which trails off like a banner in the air. The driver sits in the dead calm at the center of all this motion; hers is a lush situation."[2]

1.1
Deborah Allen in General Motors lot, Los Angeles, photographed by Oliver Allen, c. 1948. Courtesy of Oliver Allen.

1.2
Detail of "Cars '55" by Deborah Allen, *Industrial Design*, February 1955, 82. Courtesy of F&W and Oliver Allen.

1.3

"Self-portrait" of Richard
Hamilton, photographed by
Robert Freeman, used as
a wrap-around cover on
Living Arts, The Institute of
Contemporary Arts, no. 2,
1963. Courtesy of Richard
Hamilton Estate.

1.4

Spread featuring Harley Earl
Incorporated by Boulevard
Photographic Inc., *Industrial
Design*, October 1955, 78–79.
Courtesy of F&W.

1.5

Spread from "Urbane Image"
by Richard Hamilton in
Living Arts, The Institute of
Contemporary Arts, no. 2,
1963, 44–45. Courtesy of
Richard Hamilton Estate.

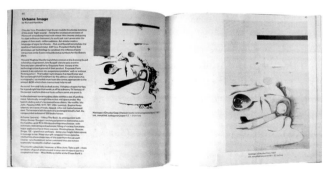

To understand the nature of a new kind of design criticism that emerged in the late 1950s and early 1960s in the US and the UK, it is revealing to look at the language and references being wielded by three writers in particular, in their attempts to reckon with the new types of mass-produced and often disposable products and appliances that began to populate the British and American marketplaces. These writers are Deborah Allen and her coeditor of *Industrial Design* magazine Jane Thompson, whom we shall meet later, and in the UK the design and architecture critic Reyner Banham and the artist and writer Richard Hamilton.

A few years after Allen's encounter with the Buick, we find Hamilton, in the library of the US Embassy on Grosvenor Square in London, leafing through back issues of *Industrial Design* and poring over its analyses of white goods and cars. Allen's phrase "hers is a lush situation" leapt from the page and became, for Hamilton, the launch point and title for a series of studies and a painting that explored the relationship between the car and the feminine form.

In 1963 he wrote an article for *Living Arts* as an accompaniment, or reference guide, to some of his art works, including "Hers is a Lush Situation" and an elaborate self-portrait he constructed for the publication's cover. His piece of writing is an ecstatic, media-fueled, and hallucinatory immersion into the imagery, accouterments, and language of the American dream. The rich accumulation of proprietary names and advertising jargon required its own lengthy glossary, in which the brands, acronyms, technical terms, oblique references, and celebrity names were decoded. These include: Virgil Exner, chief body stylist for the Chrysler Corporation from 1953 until 1961; playboy and film producer Howard Hughes; the bug-eyed monster of science fiction; Voluptua, the overtly sexual host to KABC's late-night romance movies; and technical terms and processes such as major and minor appliances, retouching, depth of field, the f2 camera stop, scanned and screened images, and dubbing.

In the text Hamilton conjures a wonder-struck aerial vision of cars driving through Manhattan and a female driver, floating, just as Allen had depicted, on waves of steel: "In slots between towering glass slabs writhes a sea of jostling metal, fabulously wrought like rocket and space probe, like lipstick sliding out of a lacquered brass sleeve, like waffle, like Jello. Passing UNO, NYC, NY, USA (point a), Sophia floats urbanely on waves of triple-dipped, infra-red-baked pressed steel."[3]

But only beats after such a sympathetic establishing shot, Hamilton taints his own imagery with the sense of an ominous and darker underside to mass production: "a prolonged breathy fart" exudes from the cars; a "comically dribbled sigh of ecstasy" is induced by manipulation of an appliance's power glide lever, likened to the Isher weapon; and the sky behind Mr Universe and Miss World is "smeared with puce and violet," for example.

Hamilton considered that in "Urbane Image" he was successful in wielding words like dabs of paint or elements of collage in order to "gauge the temperament" of the

1.6
Reyner Banham with his Moulton bicycle in Carteret
Street, London, photographed by Reginald Hugo
de Burgh Galwey in 1963. Courtesy of Architectural
Press Archive / RIBA Collections.

"A THROW-AWAY ESTHETIC"

world of American-derived appliances and branded products that he found so compelling.[4] He reflected that the article used "collage, paraphrase, style change, irony tempered with affection—a sophisticated, if superficial, erudition" to mask "a goggle-eyed wonder at the world."[5] As such, he recognized that this piece had the "ambition and character" of lyrical, even literary, prose.[6]

A third self-portrait, of a third writer under consideration in this chapter, Reyner Banham, was written on the back of a Qantas Empire Airways postcard and published in *Motif* magazine in 1963. It also sought to understand the temperament of something—the Pop movement as a whole, anthropomorphized here as a jet-setting tourist. It read:

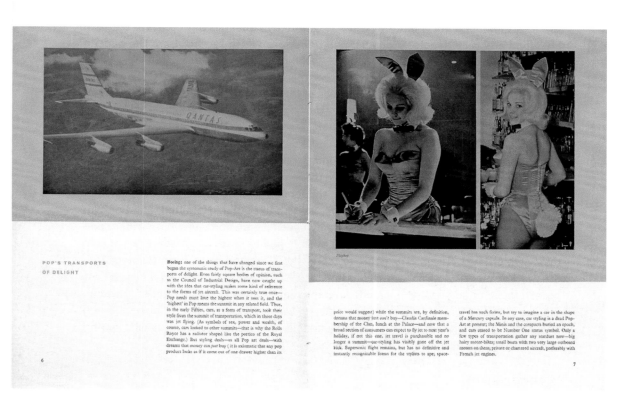

1.7
Spread from "Who Is This Pop?" by Reyner Banham, *Motif*, The Shenval Press, no. 10, 1962/3.

Dear Vidiots
Boeing along to Honolulu
—can see the islands
like Bonestells in the sea.
Now to make that Waikiki
scene, loot goodies for
home consumers. Bully for
former IG man, hey?
    Love from Pop[7]

In the article, titled "Who Is This Pop?," which follows this oblique epistle, Banham personalized Pop as both a sensibility and a kind of father figure, and in doing so positioned himself as the phenomenon's paternal representative and transla-tor of its unfamiliar vocabulary. The seven component qualities of Pop which Banham identified embrace people, products, and behaviors which "spring from the sudden collision of a keen far-out taste and the technical means to satisfy it."[8] The piece contributed a new understanding of the consumers of popular products, not as passive dupes of commerce, as other cultural critics of the period might have considered them, but rather as sophisticated and knowing participants in what Banham termed "the extraordinary adventure of mass production."[9]

In the post-Second World War period, the dominant strain of design proselytizing, typified by the activities of the Council of Industrial Design in the UK, and the Museum of Modern Art in the US, evaluated design using abstract and morally value-laden cri-teria such as durability, honesty, sobriety, and modesty, and was elitist in intention and omniscient in voice. Against this backdrop, writers like Allen, Hamilton, and Banham wanted to highlight the ways people actually used design, and allow for a wider spec-trum of consumer needs and tastes; and even hoped to empower readers to conduct their own product design criticism—to "carry the discipline down from Olympus into the market-place," as Banham would later put it.[10] These writers used a direct, first-person mode of address, and included personal anecdotes and experiences. Through the use of such literary devices as neologisms, compression, rhythmic play, and rich imagery, a new genre of writing gained definition, transcending its journalistic setting, and aspiring to a hybrid form of poetic prose.

INTRODUCTION

The witch hunt for design's "sensational aspects"

British design commentary in the mid-twentieth century was, for the most part, pro-motional and didactic—a form of economically driven and government-endorsed public service intended to improve the quality of British design, the taste of British retailers and consumers, and the profitability of British manufacturing. Its language, references, and philosophical underpinnings, rooted in the cleansing rhetoric of mid-nineteenth-century design reform, had been reinforced through the publications, broadcasts, and exhibitions of early twentieth-century institutions.

Similarly, in the US in the 1950s the discussion of design was linked to efforts to boost the sales of American industrial design in the marketplace. Such discussion was led, for the most part, not by government bodies, however, but by an alliance of cultural institutions, publishing, and retail businesses. At the forefront was the Museum of Modern Art, which staged multiple exhibitions and competitions as part of its "Good Design" program, directed by Edgar Kaufmann Jr. Between 1950 and 1955 three exhibitions were staged each year, two at the Merchandise Mart in Chicago and a third at the MoMA in New York, with award-winning examples of good design indicated with distinctive orange and black labels.[11] The Museum's activities were further promulgated through its own publications, talks and panel discussions, and through lengthy reviews and transcripts of these events in trade architecture and design publications. Here the emphasis was on commercial competitiveness, what *Interiors* magazine summarized as a combination of "facility and economy of manufacture, and sales appeal," and what the selection committee of the 1951 "Good Design" exhibitions described as "a real contribution, in looks, in efficiency or in price."[12]

Whether they originated in the UK or the US, definitions of what constituted "good design" or "contemporary design" often sounded similar. Each referenced Arts and Crafts and Modernist-derived moral and aesthetic values, which advocated that the structure, means of manufacture, construction materials, and purpose of a product should all be evident, while decoration should not.[13] Kaufmann asserted that "Modern design should be simple, its structure, evident in its appearance, avoiding extraneous enrichment." His "Good Design" selection committee passed over "pieces that would dominate a room by their sensational aspects" in favor of ones "that showed a more controlled design."[14] In the UK, Paul Reilly, Head of the Council of Industrial Design's Information Division, echoed this preference for restraint and this tendency to transpose human character traits onto furniture. He specified qualities such as honesty, decency, straightforwardness, and modesty in order to distinguish "contemporary design" from "the rootless, vulgar, modernistical furniture that glittered in chain store windows."[15]

The new design critics discussed in this chapter were often hired to write for and about institutions like MoMA and CoID, but, as independent critics, they were not beholden to them. Richard Hamilton, for example, referred to MoMA as "a custodian of relics as well as a propaganda machine," while Reyner Banham castigated CoID for its "narrow, middle-class" interpretation of taste, and its misguided belief that there was "some kind of necessary relationship between the appearance of an object and its performance or quality."[16] These writers questioned the CoID's official line on contemporary design and the

1.8
Cover of *Collected Words* by Richard Hamilton,
showing some of the design publications he
contributed to in the late 1950s. *Collected Words*
by Richard Hamilton (London: Thames & Hudson,
1983). Courtesy of Richard Hamilton Estate.

normative rhetoric of MoMA's "Good Design" program, demonstrating, instead, an appreciation of surfaces, symbols, and styling, technological advances, planned obsolescence, and the perspective of the knowing user-consumer.

American Studies historian Daniel Horowitz observes that in the 1960s new types of writing about consumer culture emerged in Western Europe and the US that encapsulated changing attitudes toward pleasure and playfulness.[17] Acknowledging that early-twentieth-century writers had also dealt with pleasurable experiences created by commercialism, but had usually "linked them with what they considered lowly, corrupting and escapist indulgences such as excessive drinking and illicit sex," Horowitz posits that by the 1960s changing moral attitudes had allowed for new ways of looking at consumer culture.[18] He identifies the ways in which writers such as Tom Wolfe, Umberto Eco, and Roland Barthes challenged the divide between high and low, adopting what he calls an "anthropological outlook" on culture.[19] To Horowitz, these writers were "increasingly focused on pleasure, playfulness, and sexuality as key aspects of a more positive interpretation of commercial culture. They wrote of the way automobiles, clothing, the built environment, comics, advertisements and movies enabled people to gain emotional enrichment from commercial goods and experiences."[20]

Such writers depicted consumer culture as a broadly defined social phenomenon, and its products typologically rather than specifically. While Wolfe tended to use designed objects in his writing as stage props to support the veracity of his detailed character portraits, and Eco and Barthes studied them in essentialist terms, the design critics brought to light in this chapter engaged more directly with the design, manufacture, and use of commercial goods. Writers like Banham and Allen used poetic language to illuminate the products they depicted, rather than using the names of products to enliven their prose, and addressed the detail of specific year models and editions rather than generic types.

A different, and more anxious, current of criticism also gained traction in the latter half of the 1950s, particularly in the US. This directed public attention toward the adverse effects of the product design industry on the environment and on society, and was written for a general audience by commentators such as the sociologist C. Wright Mills, historian Daniel Boorstin, and the lawyer and author Ralph Nader. Such writers considered consumer goods from the perspectives of fields beyond design such as sociology, economics, politics, and ecology. By the early 1960s these two very different forces of resistance—literary, on the one hand, and sociological, on the other—were beginning to disturb, and in some cases redirect, how and why interpretative commentary about design was conducted, and whom it was for.

This chapter charts the emergence and impact of a genre of writing, attuned to its role as criticism, that represented a new attitude toward the design, manufacture, and use of consumer products in the postwar period in the US and the UK. This new writing countered the elitist and didactic motivations of official design propaganda through interpretations that embraced emotional responses as well as practical realities, and through the introduction of antiestablishment values. Most notably, this new brand of writing aimed to engage readers through the use of heightened language which can be termed literary product design criticism.

PART ONE: A CLASH OF VALUES IN *DESIGN* MAGAZINE, BRITAIN, 1960

A "pre-occupation with honesty in design"[21]

In the February 1960 issue of *Design*, the monthly journal of the British Council of Industrial Design (CoID), readers found an article that did not seem to belong with the journal's typical content. Titled "Persuading Image," it was written by the artist Richard Hamilton, who, in the late 1950s and 1960s, was gaining recognition as one of the founders of the British Pop Art movement but was also practicing as a designer, teaching in the Royal College of Art's interior design department and the fine art department at King's College, Newcastle, and writing about design in *Architectural Design*, the *Architects' Journal*, *Uppercase*, and *Design*, among other publications.

In "Persuading Image," Hamilton wrote about how, during the 1950s, American industrial manufacturers and designers had used sophisticated and witty imagery to seduce their consumers, to "mould" them to fit the products they had already created, and the implications of these precedents for manufacturing, marketing, and consumer practice in 1960s Britain. Hamilton's positive interpretation of these calculating industrial practices, and his serious consideration of such issues as styling, image retouching, motivational research, and planned obsolescence, disrupted *Design*'s narrow editorial perspective, visually, tonally, and in terms of its values. *Design*'s philosophy was based upon "well established principles," which John Blake summarized in his editorial preface to the February 1960 issue as "truth to materials, to production techniques, to the expression of the nature of a product and its function and, more recently perhaps, to the fulfilment of basic human needs."[22] According to Blake, who used an anecdote in which "an American designer recently expressed bewilderment at his young British assistant's pre-occupation with honesty in design," there was a profound disjuncture between British and American views of design's positioning in society.[23] While the British considered design a social and moral concern, Blake and *Design* magazine averred, the Americans, apparently, could conceive of its value only in commercial terms.

Hamilton's idiosyncratic take on the social benefits of advanced capitalist product design challenged the established viewpoint of the CoID, and the British design professions it represented. His article offered a more pragmatic, style-oriented, and American-influenced perspective on the inevitability of capitalism and the designer's complicated role therein.

A "duty" to fight against "shoddy design"[24]

In order to understand the ways in which Hamilton's article jarred with the ethos of its host publication, it is necessary to take a look back at the formation of these values. The Council of Industrial Design was established in London in 1944 by the Coalition Government in anticipation of the need for a postwar boost to Britain's manufacturing industries, and to help the transition from the state-controlled production of wartime to a mixed market-based system.[25] The CoID translated Britain's need for a competitive edge in international markets into a two-pronged domestically focused mission: to raise consumer taste and to encourage manufacturers to produce better-designed goods. In its campaign to raise standards as a matter of civic duty, the CoID followed in the footsteps of other propagandizing organizations established in the first half of the twentieth century, such as the Design and Industries Association (1915), the Council for Art and Industry (1933), the Society of Industrial Artists (1930), and the Royal Designers for Industry (1936).

The CoID was funded by the Board of Trade, and while the main impetus for improving design standards was economic, the CoID was also a direct descendant of the nineteenth-century design reformists who believed in the power of good design to effect *social* change and to uphold *moral* values.

Nineteenth-century design reformists transposed moral virtues to the field of craft production, invoking such tenets as "truth to materials" and "honesty of construction" in their efforts to improve the aesthetic quality of the decorative arts and the condition of the society in which they were produced. John Ruskin, a leading critic of the Victorian era, saw the state of decorative arts and architecture as indices of the spiritual health of society. He was concerned that Britain's too-rapid industrialization would obliterate its natural landscape with mills, quarries, kilns, coal pits, and brick fields. In a lecture at the Bradford School of Design in northern England, he said: "Unless you provide some elements of beauty for your workmen to be surrounded by, you will find that no elements of beauty can be invented by them."[26]

William Morris, a socialist writer and designer who became the best-known theorist of the late-nineteenth-century Arts and Crafts movement, also linked aesthetics to social conditions. In a lecture in Burslem, a town at the center of the Midlands pottery industry, he spoke of the correlation between a beautiful living environment and the creation of beautiful design, conscious that as he spoke he was standing "in a district that makes as much smoke as pottery."[27] He set the notions of "art" and "dirt" in opposition, suggesting that the land would have to be turned from the "grimy back yard of a workshop" into a "garden" in order for art to flourish: "Of all the things that are likely to give us back popular art in England, the cleaning of England is the first and the most necessary."[28]

The reformist mission to tidy up England and to cure its social ills had a pervasive legacy due to the widespread distribution of Morris and Ruskin's published writings and the design-related institutions they helped to shape, and they still informed the tenor of

1.9
Advertisement for The Design Centre, London,
emphasizing its close affinity to retail, in
Design, August 1960, 18. Courtesy of University
of Brighton Design Archives.

most CoID activities more than half a century later. For example, a 1936 article charting twenty-one years of the Designers and Industries Association devoted a spread to a family tree of influences converging on current DIA exhibitions and publications, and at the head of this family tree was a photograph of William Morris.[29] Additionally, Gillian Naylor, an editorial assistant at *Design*, has recalled the importance to the CoID of William Morris specifically, and that "once, in an editorial, C. R. Ashbee was spelt Ashby and Gordon Russell, then director of the CoID, pointed this out and said, 'These are the people this institution is founded upon and you must at the very least get their names right in the magazine!'"[30] Echoes of Morris and Ruskin could be heard in architecture and design critic Nikolaus Pevsner's assertion, in a 1946 BBC Radio broadcast, that "Bad design is just as devastating for people as bad air and over-long hours."[31] In the same year, design critic Anthony Bertram, writing about the "Enemies of Design," was hopeful that a group of new reformers he called "the Doctors" would succeed through their writing in "trying to heal the present tragic disease of the useful arts."[32]

One of these "doctors" and a keeper of the reformist flame was Michael Farr, who edited the CoID's journal *Design* from 1952 to 1959, and stayed on as Director of Information until 1962. While studying English Literature at Cambridge, with the literary critic F. R. Leavis, he met Pevsner, who was then the Slade Professor of Fine Art at the university.[33] Pevsner asked Farr to write the revised edition of his 1937 tome, *An Inquiry into Industrial Art in England*, and so Farr's introduction to design was a kind of trial by fire in the form of a rigorous study of hundreds of British manufacturers, designers, and retailers, conducted under Pevsner's supervision.[34] In the introduction to *Design in British Industry: A Mid-Century Survey*, Farr revealed his belief—in line with Pevsner's, and channeling those of Morris and Ruskin—that the mission of design reform was inextricably connected to that of social reform:

> One cannot approve of thoughtless and insensitive designs. Neither can one approve of dishonest designs, such as a pressed glass bowl trying to look like cut crystal glass, a plastic-covered handbag made to resemble snakeskin, an aluminium tea-pot masquerading as hand-beaten pewter. In the same way one cannot approve of imitations of period designs. As we shall see, the arbitrary invocation of antique styles is a disease from which few industries are free. ... Such false and meretricious designs attempt to provide a substitute for the needed splendour which all aspects of our environment should be made to concede. The pleasure which most people take in an entertainment so vicarious as the cinema, as well as the pleasure in vulgar and boastful design, is largely accounted for by the universal longing to escape. Looked at from this point of view, the question of industrial art is a social question, it is an integral part of *the* social question of our time. To fight against the shoddy design of those goods by which most of our fellow-men are surrounded becomes a duty.[35]

Farr depicted a designed landscape infected by mass culture and such antisocial values as "thoughtlessness," "insensitivity," "dishonesty," "falsehood," "vulgarity," "boastfulness," and "shoddiness," all of which he believed it was his "duty" as a design critic to "fight against." Through enumerating the evils of an environment lacking in "splendour," he conjured a conception of a contrasting ideal society, guided by the direct opposites of such values, namely: honesty, functionality, taste, modesty, and craftsmanship and durability. The ideals expressed in this passage, inherited from the design reformist tradition, as well as the use of the formal, seemingly objective third-person pronoun to express strongly subjective and elitist views, were typical of the content of the numerous publications issued by the CoID well into the 1960s.

"One more word about teapots": *Design* magazine[36]

Design magazine, founded in 1949 as the CoID's journal of record, functioned as another weapon in the Council's propagandist armory, alongside its exhibitions of good design held at the Design Centre in Haymarket, London (opened in 1956), its Good Design Award Scheme (begun in 1957), its educational films, wall cards, portable box exhibitions, newsletter, joint ventures with the BBC and Penguin Books, and its Design Index (a catalog of British products which met the Council's selection criteria). The CoID was organized into the Industrial Division and the Information Division, which corresponded to its dual objective of "the creation of a Supply of good design" and "the creation of a Demand for good design," respectively.[37] The magazine, which fell under the auspices of the Information Division, reported on consumer goods already endorsed by the Council; as Gillian Naylor remembers it, "The CoID used to feed us material which they wanted us to feature in the magazine."[38] The schematic organization of the magazine also derived from the CoID Design Centre, which grouped its objects and design files "as far as possible to correspond with department store practice."[39]

The CoID's main role was to define and extol "good" design in order to try and persuade the British public that modern design was what people should be buying.[40] "Appropriate materials," "good appearance," "good workmanship," "suitability for purpose," and "pleasure in use" were some of the recurring criteria by which products were selected for the Council's Design Index and, by extrapolation, for inclusion in the magazine. A 1954 Readership Survey revealed that *Design*'s readers were not particularly inspired by the magazine's reliance on the unexplained absolute of "good design" for its editorial decision-making. The survey makers summarized the readers' responses by saying: "There is a good deal of demand for articles presenting points of view other than an 'official' one."[41] It is also telling that the survey found that the most popular section in a magazine devoted to the improvement of British product design was the "Foreign Review."[42]

The readership survey also reveals that very few members of the so-called "British public," beyond designers, were actually reading *Design*, and it was available only at specialist bookstores and through subscription. Of its 13,600 readers in 1960, the highest percentage worked in the furniture and appliances sector, and the second-highest readership was in education.[43] And yet, since few other British publications of the period were focused solely on contemporary industrial and product design, the journal filled a significant gap in the market for coverage of its subject. Its editors aspired to emulate the *Architectural Review*, which at the time was commissioning thoughtful and much-debated examinations of urban concerns by such outspoken and incisive writers as Ian Nairn, whose "Outrage" and "Counterattack" features on the spread of urban sprawl in the UK garnered attention for the *Review*. In reality, *Design*'s nearest competitor, for a general audience at least, was probably *Which* magazine, the journal of the Consumers' Association, a product-testing and consumer advocacy charity founded in 1957, which by 1959 had 150,000 members. The small 32-page magazine included reports of its laboratory testing on electric kettles, aspirin, cake mixes, scouring powders, non-iron cottons, and the like. *Design* stated that its intention was to evaluate design based on an analysis of the distance between "the promise of the product's appearance" and its actual ease of use, but was understandably keen to distinguish itself from *Which*, whose February 1960 issue featured hormone creams, cycle rear lamps, and deodorants—products that rendered *Design*'s evaluative criterion somewhat futile.[44] *Design* reaffirmed that its intention, therefore, was "less to provide a guide to what is best on the market than to suggest, through a close study of individual products, what are the things that really matter in design, and consequently what are those areas of investigation and research which are of most concern to designers and manufacturers."[45]

The other articles in the February 1960 issue of *Design*, in which Hamilton's unorthodox article was published, included a piece on street furniture with an introduction by the Minister of Transport; an article about a stool designed to help factory workers move between work stations; and a report about design in Czechoslovakia. These kinds of informative, but rather dry, articles about aspects of design's application to British industry were a mainstay of the magazine at the time. As John Blake recalled, "Much attention was devoted to ergonomics, then emerging as a practical science that could describe more precisely the physical and psychological needs of people."[46] The showcased projects tended to be worthy municipal initiatives, evaluated in earnest, sometimes hectoring language, and illustrated with black-and-white photographs. Bruce Archer's analysis of a new range of melamine cups and saucers represents the extreme of the quasi-scientific approach to design evaluation that pervaded the magazine by the early 1960s. Archer, who sought to bring the robustness of his engineering background to bear upon design criticism, started a series titled "Design Analysis" in 1957. He took on one product every other month, and used a set of concrete standards he had developed in order to measure the worth of its design. "By selecting one product at a time," Farr reflected of Archer's project, "it shows how the design stands up to technical cross-examination at the manufacturing stages, and functional analysis at the point of use."[47]

DESIGN ANALYSIS 17

Cup and saucer in melamine

L. BRUCE ARCHER

DESIGNER *Ronald E. Brookes*. MAKER *Brookes and Adams Ltd.*
PRICE *including tax (cup and saucer) 6s 6d (single colour) ; 8s 3d (two colour).*

The melamine cups and saucers from the new Fiesta range of tableware by Brookes and Adams Ltd, represent a break from traditional designs. This same firm introduced some of the first plastics tableware to Britain as long ago as 1923, when urea was the material used.
In this analysis the author outlines the manufacturing problems involved with plastics cups and saucers and describes the way in which their solution has led to the new designs. Opinions of users in seven households are taken into account in the author's assessment. The manufacturer's comments are on page 47.

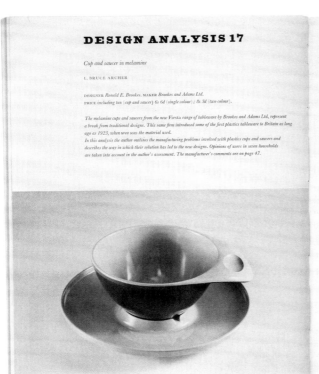

The stigma of being regarded as a cheap substitute for earthenware lies heavily upon plastics when used for tableware. The echo of that indictment, "cheap substitute", remains at the back of almost everybody's mind, and persists even though modern plastics materials such as melamine not only have some advantages over earthenware, but are also (substantially) dearer. The manufacturer who wishes to increase his sales of plastics tableware is therefore faced with a difficult marketing problem.

Fundamentally, the problem is how to get plastics tableware out of the picnic and kitchenware departments of the retail stores and into the tableware departments, which are usually called 'china and glass'. The picnic and kitchen market is both limited and price sensitive, but at the same time china and glass salesmen are snobbish about plastics. However, according to *Industrial Design* (November 1958 page 96) more than 50 per cent of all dinnerware sold in the United States is now in melamine, so the problem is obviously not insoluble. This then, was the background against which Brookes and Adams embarked on the new *Fiesta* designs.

Objectives Two features which had hitherto kept plastics out of china and glass departments, despite melamine's considerable functional advantages, were its monotone colouring and its unsympathetic feel.

An obvious approach to the problem of decoration was two-tone moulding — an established technique, but one which normally entails even higher manufacturing costs. The manufacturing difficulties are of two kinds.

Firstly, a conventional loop handle requires moving cores in the mould which must be retracted from the loop before the finished product can be removed from the die. Dies with moving cores are not only expensive to make but they are also subject to more rapid wear, and there is a greater danger of production stoppages due to mechanical faults. Output is slow and there is a higher proportion of rejects. Moreover, it is difficult and expensive to eliminate flashlines on the cup around the handle, where the cores slide into the die. Secondly, a two-tone product is manufactured in two stages, the first colour being moulded by an oversized punch in the final die, thus producing a thin-walled product. The additional colour is moulded on to the first in the same die by means of the normal punch. The chief difficulty here is to ensure that both colours are cured to the same degree and truly bonded, even though one is in the mould longer than the other.

The solution The first step towards solving the marketing problem was to emphasise that melamine is a material in its own right with qualities which make it ideally suitable for table use — durable, hygienic, and reproducible with accuracy in both detail and colour. It should have been obvious a long time ago that the industry's habit of copying shapes, such as the traditional handle, which have arisen out of the nature of ceramic materials, was not only bad design for plastics production, but was also the mark of the substitute material.

The next stage was to decide to present the product in display packages which would have a better chance of being shown in tableware departments than would naked plastics cups amongst china and glass. Having chosen to design cups and saucers which would not snugly in a

DESIGN 134   45

1.10
"Design Analysis" by Bruce Archer, *Design*, February 1960, 44–45. Courtesy of University of Brighton Design Archives.

The left-wing poet, essayist, and former editor of *Encounter* magazine Stephen Spender, in a 1958 speech to the Society of Industrial Artists, crystallized mounting unease among the design-conscious public about CoID's stultifying bias toward functionalism. Spender saw too many designed goods "pincered" between the "two extremes of utilitarian functionalism—the aeroplane on one flank and the kitchen utensil on the other."[48] He listed the visual attributes of functionalism as "bareness, simplicity, squareness or roundness, solidity, seriousness," and warned that equating functionalism with beauty (a message often contained in the pages of *Design* magazine) was really a sleight of hand. He continued:

> I know the objection to my way of thinking. It is that designers are designing today for socialised welfare state man, leading him down the Welwyn Garden path, educating him gently with discourse piped from the Third Programme. None must talk too loud, no one must flash a light too brightly in his eyes, there must be no violent splashes of colour, he must be anaesthetised with good taste, and who but the British, with the British Council, the Arts Council, the Third Programme, the Design Centre, panethol, chlorophyll, Dettol, know most about disinfectants and anaesthetics.[49]

A challenge to the taste anesthetists

Only a year later, along came a writer more than willing to talk loudly, flash lights, and splash color in the faces of the taste anesthetists at the CoID. With his "Persuading Image" article, Richard Hamilton upset the delicate balance of good taste, belief in the conflation of usefulness and beauty, and adherence to design reform social values that the CoID had endeavored to maintain in the postwar years.

Hamilton was born in London, the son of a car showroom driver, and disparately schooled in art at a variety of adult education evening classes, the Royal Academy Schools and, when they closed in 1940, in engineering draftsmanship at a Government Training Centre, and finally at the Slade. In the late 1950s, Hamilton was developing an art practice inspired by popular culture and a writing practice through which he tested his ideas. Recalling his life at the time, he said: "Why was I going to the cinema three times a week, and reading *Esquire* and *Life* magazine and then going home to the studio and painting monochrome squares and hard-edged abstraction? It didn't seem to fit. So I tried to incorporate the material I was interested in—the sociological aspects of current living—and create a kind of aesthetic which would enable me to produce a painting that I felt reflected the situation in which I found myself. Writing helped me work through these ideas."[50]

The "Persuading Image" article was based on a lecture that Hamilton had delivered in 1959 at the Institute of Contemporary Arts (ICA) to members of the Independent Group. This loose-knit salon included the artists, critics, and architects Lawrence Alloway, Reyner

Banham, Theo Crosby, and Alison and Peter Smithson. The group had been meeting since 1952 to plan exhibitions and discuss ideas about the machine aesthetic, science fiction, communication theory, and other aspects of Pop culture, specifically American Pop culture, as a rebuttal to prevailing standards of good taste in 1950s Britain, and in particular those of the founders of the ICA, the critics and collectors of modern art Herbert Read and Roland Penrose.[51]

The Independent Group discussions were motivated by an impulse to break down the divide between high and low culture. Group members prided themselves on being genuinely interested in, and bona fide childhood consumers of, what Banham termed "the popular arts of motorized, mechanized cultures ... like the cinema, picture magazines, science fiction, comic books, radio television, dance music, sport."[52] Alloway, in particular, theorized this position in relation to the popular arts. In his 1959 article "The Long Front of Culture," he presented a conceptual model that put culture along a horizontal spectrum, rather than stacked in a hierarchical pyramid, with mass culture at the bottom and refined high culture at the top: "unique oil paintings and highly personal poems as well as mass-distributed films and group-aimed magazines can be placed within a continuum rather than frozen in layers in a pyramid."[53] Alloway's article dismantled the idea that the arts were the exclusive possession of an elite, and argued that permanence and uniqueness were not the only criteria by which the value of material culture might be judged.

Among the influential exhibitions the Independent Group organized, with Alloway's premise at their centers, were: "Man, Machine and Motion" (ICA, 1955), in which Hamilton attached blown-up photographs of machines in use to a modular steel frame; and "This Is Tomorrow" (Whitechapel Gallery, 1956), in which Crosby coordinated twelve teams of artists, critics, and architects who each explored through mixed media different aspects of the future. Hamilton's poster for this exhibition, titled "Just what is it that makes today's homes so different, so appealing?," was a collage of images sampled from magazines, including a 1955 Constellation Hoover, a tin of Armour Star Ham, and a photograph of the earth, which Hamilton saw as representative examples of the categories of emergent Pop iconography, such as domestic appliances, processed food, and space.

The Independent Group also arranged a series of lectures for small groups of invited guests that included expositions on Elvis and on violence in the cinema. Hamilton's lecture, "The Design Image of the Fifties," delivered in the fall of 1959, investigated domestic appliances such as washing machines, vacuum cleaners, radios, and refrigerators, and the role of advertising in creating the image of these consumer goods.[54] He described his ICA presentation as being rather "exotic." He had three projectors and three screens, one of which took up the whole of the back wall of the ICA meeting room, then located at 17–18 Dover Street. This format was Hamilton's response to the multiscreen film *Glimpses of the U.S.A.*, produced by the American designers Charles and Ray Eames for the 1959 American National Exhibition in Moscow, in which 2,200 images were projected on seven twenty-by-thirty-foot screens, and which had been reported on in the April 1959 issue of *Industrial Design* magazine. "In my modest little way I was trying to catch up with the avant-garde," Hamilton recalled.[55]

How to design a consumer for a product

Hamilton's lecture and article described how a sophisticated image industry had constructed the internationally exported image of an opulent 1950s America.[56] He examined the relationship between designers, manufacturers, publicists, magazine editors, and consumers, and how the image functioned at each juncture of the production, distribution, and consumption of goods. He was particularly impressed by the ways in which American manufacturers hired image makers to manipulate consumers to buy the products they had already created: an efficient system, in his view, where "the consumer can come from the same drawing board" as the product.[57]

Hamilton argued that while British designers were "interested in the form of the object for its own sake as a solution to given engineering and design problems," social and economic realities had, in fact, effected a complete reversal of these values. What American designers realized, he thought, was that the most important aspect of design was not appearance or usefulness, but rather the maintainability of the production and consumption cycle. Hamilton recounted how American designers had developed "a new respect for the ability of big business to raise living standards," and big business now appreciated "the part that design has to play in sales promotion."[58] In Hamilton's view, the virtue of American industrial design was that it had "come to terms with a mass society" in ways that British designers still seemed incapable of. Functionality now had to encompass how well a product was working in the market.

Hamilton's article was conspicuous in the pages of *Design* through its role as a conduit for American perspectives on economic and social practices. The idea, for example, that manufacturing efficiency and national prosperity were contingent upon accelerated obsolescence had been propounded by American economists like Peter Drucker.[59] The sentiment that "we are obligated to work on obsolescence as our contribution to a healthy, growing society" was typical of free-market economic thinking in late-1950s America.[60] Hamilton had also read the articles of American industrial designer George Nelson, who, in his 1956 article on "Obsolescence," had explained that America's wealth was dependent on its wastefulness, which enabled mass production at an ever-increasing pace, and "provides a way of getting a maximum of goods to a maximum of people."[61] Nelson assumed that the European view of this situation would be "a blend of appalled curiosity, downright disbelief, righteous indignation and envy."[62] Hamilton's response defied this expectation; he was convinced that "increased productive capacity is a basic social good."[63] And, in fact, the righteously indignant response to the issue came from an American, the social critic Vance Packard, whose book *The Waste Makers*, a hard-hitting social critique of planned obsolescence, was published in 1960 and crystallized concern over the contribution of planned obsolescence to a perceived crisis in American cultural values.

For Hamilton, however, an appreciation of "big business" (a term he had previously brandished in his 1957 enumeration of the qualities of the emerging genre of Pop Art) and mass society was what was missing from British industrial design.[64] "Even the production of goods of dubious value, is, in the long term, likely to benefit society," he argued.[65] "Change is most likely to occur in those objects that least deserve to live," he opined in another article about obsolescence.[66] He was clearly frustrated by the slow uptake of such ideas on his side of the Atlantic.

Most other articles published in *Design* betrayed a bias toward the social utopian ideals they saw as materialized in pure forms dictated by function, natural materials, craftsmanship, and the work of Scandinavian designers, for example. Hamilton, by contrast, provided a glimpse of the economic reality in which mass-produced design actually operated. He conceived of industrial designers not as craftspeople but as canny commercial operators, describing them variously as "marketing aids," "men who establish the visual criteria," operators of "the machinery of motivation control," and collaborators with "ad-man, copywriter and feature editor."[67]

Hamilton admired the people who knowingly constructed the "designed image of our present society" in the pages of "glossy magazines." These were the very images, after all, that provided him and other members of the Independent Group with such a rich source of raw material for their discussions and artwork. He talked of their creators' "skill and imagination" and "wit," and quoted their slogans—"plush at popular prices"—surely aware of the goading effects such language would have on *Design* magazine's editors and readership.[68] Even the use of the single word "glossy" would have triggered complex reactions in postwar Britain. Paper rationing continued in Britain for several years after the end of the war, and even by 1960 British magazines were rarely printed on gloss paper. Hamilton was using the word to describe American mainstream magazines such as *Life*, *Look*, and *Esquire*. But he was also aware of the pejorative nature of the term's metaphoric connotations in a postwar Britain fearful of the Americanizing influences of "ersatz" and "candy-floss" mass arts on the previously "organic" expressions and "oral traditions" of working-class culture, as left-wing sociologist Richard Hoggart had termed them.[69] The 1961 *Design* Readership report, produced after Hamilton's article had been published, revealed that some readers considered that even *Design* was becoming too glossy. The director of a firm producing tubular steel products opined: "There is too much window-dressing by art people and it has gone off the functional idea. It is tending to become an art-glossy."[70] In his 1957 book *The Uses of Literacy*, Hoggart used the term "glossy" as a negative label for the kinds of furniture shops, novelettes, and magazines he believed were exerting such a worrying influence on British society. In describing monthly pin-up magazines, he wrote: "The 'cheesecake' is a little more advanced than most newspapers would be prepared to print at present, and especially well photographed on glossy paper."[71]

        "A THROW-AWAY ESTHETIC"

Hamilton was keen to draw a distinction between the popular arts, "in the old sense of arising from the masses," as championed by Hoggart and others, such as the author George Orwell, and his own conception of a more industrialized and calculated Pop Art, which he saw as stemming from a professional group with a highly developed cultural sensibility. In a 1960 lecture at the National Union of Teachers conference, Hamilton reemphasized the difference he perceived between unsophisticated working-class popular arts such as club singing, on the one hand, and the current manifestations of the commercially driven, urbane popular culture he was so fascinated by, on the other: "The analysts of popular culture in recent years have been negative in their approach. Whyte, Packard and Hoggart, whose ideas as we know have been given full rein in the mass media, are unanimous in their condemnation [of 'gloss, glamour and professionalism']. The story is the same: the end of the world is upon us unless we purge ourselves of the evils of soft living and reject the drive for social and economic advantages."[72]

The title of Hamilton's *Design* article, "Persuading Image," evokes the title of another of Vance Packard's books, *The Hidden Persuaders*, a best-selling critique of motivation research (MR), a practice being used by the American advertising and marketing industries to "depth-probe" the consumer psyche. Based on methods used by the government during World War II, which drew on the depth psychology of Freud, but also on sociological and anthropological research techniques, MR attempted to ascertain the effects of consumers' psychological weaknesses on their buying habits. Packard identified eight "compelling needs," including secret hostilities, guilty feelings, and sexual impulses, that marketers convinced people they might fulfill through the products they bought. "These depth manipulators are, in their operations, beneath the surface of conscious life, starting to acquire a power of persuasion that is becoming a matter of justifiable public scrutiny," he wrote.[73] Hamilton thought that "the effect of this criticism of our culture, coloured as it is by the hysterical overtones of its re-interpretation within the mass media, has been to create an atmosphere of unrest, which can itself be dangerous."[74] His use of the word "Persuading" in his title invoked Packard's work, therefore, but denied its moral viewpoints.

A definition of the term "affirmative" in the glossary of the "Urban Image" piece Hamilton wrote for *Living Arts* in 1963 gives some insight into his critical stance. Of his choice of the word "affirmative," Hamilton wrote: "Yes. Somewhat forced expression of need to conclude on a grandly positive rhetorical note. An art of affirmatory intention isn't necessarily uncritical; though I affirm that, in the context of our present culture, it will be non-Aristotelian."[75] For Hamilton, this idea of non-Aristotelian logic, in which value judgments were irrelevant, derived from an influential conversation with Lawrence Alloway about the physicist John von Neumann, whose work on quantum logic suggested to Hamilton that "we can't take a moral position any more because it's all to do with flipping coins and roulette wheels and chance, and that means there is no longer any justification for Aristotelian logic."[76]

Hamilton's use of the word "Image" in his title was probably equally provocative at the time. At *Design*, there was a wariness toward the visual image; the technical and functional attributes of products were always stressed, as if in compensation for the superficial allure of their visual appearance. In a review of two graphic design exhibitions held in London in 1960, for example, John Blake observed of an image's need to compete for attention with its neighbor: "there is a danger in such demands for attention, for the designer is tempted either to produce work that is vulgar or, in escaping from this, to resort to sophisticated pattern-making."[77] Suspicion of the image in *Design* tended to be conflated with suspicion of American culture and design. Referring to an exhibition of American design held at the US Trade Center in London, Blake derided the commercial nature of American packaging design, concluding that there was little on display "that would have been acceptable to even the most catholic of British selection panels."[78] He added that the difference between a British and an American designer lay in the fact that the latter was "untroubled by the pangs of conscience that afflict at least some of his European colleagues."[79] This resentment of America and its popular culture exports, which were a blatant reminder of Britain's comparative economic decline after the war, was widespread in British social commentary of the period. American values were often used in the postwar years as a new target to replace Fascism in the public imagination. Cultural critic Raymond Williams wrote in 1962 that the very worst of the mass media "is American in origin. At certain levels we are culturally an American colony. ... To go pseudo-American is a way out of the English complex of class and culture, but of course it solves nothing; it merely ritualises the emptiness and despair."[80] This kind of critique was not confined to the UK; in the US, too, there was rising concern about the social symptoms of mass consumerism. In his 1961 book *The Image*, historian Daniel Boorstin, for example, defined an illusion as an "image we have mistaken for reality," and observed ruefully that "The making of illusions which flood our experience has become the business of America."[81]

For Hamilton, however, as a fan of America and its "illusion" industries of advertising and public relations, the constructed image was far from sinister. The exchange of art-directed images was a shared connection among his peers, who were obsessed by what he termed "pin-board culture." "We would walk into the houses of friends and find we all had exactly the same picture," he said. "There was a picture of an American model wearing a backless dress, showing her backside cleavage, which was very venturesome. That was on everyone's pin boards."[82] Hamilton used the image of swimwear model Vikky Dougan's back, framed by a low-cut white dress, clipped from a 1957 *Esquire* magazine feature, to illustrate "Persuading Image," as well as for the artwork he created between 1958 and 1961 titled "$he."

Hamilton's fascination with the art direction and construction of images is also evident in his inclusion of the 14-item list of the crew and props it took to produce the complex self-portrait that was used for the cover of the ICA publication *Living Arts*, discussed above. The image features a 1963 Ford Thunderbird with a lingerie model sprawled on the back, and Hamilton himself wearing an American football uniform leaning on the hood, a Mercury spacecraft capsule on loan from Shepperton Film Studios, a refrigerator stuffed with American food, a Wondergram, a vacuum cleaner, telephone, typewriter, and toaster, all arranged on a background of high-gloss pink paper on the pavement outside the Taylor Woodrow building in West London, and photographed from an upper floor by Robert Freeman. Hamilton credited the photographer and the stylist, and himself as the "producer" of the image.

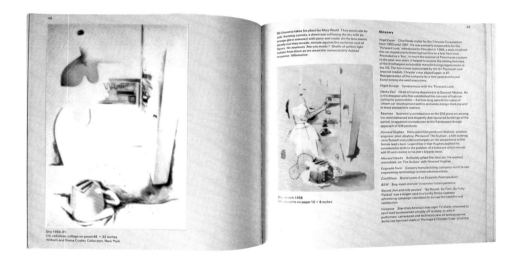

1.11
Spread from "Urbane Image" article by Richard Hamilton, including a painting and sketch in the "$he" series, based on domestic appliances from *Industrial Design* magazine and the image of Vikky Dougan's back. "Urbane Image" by Richard Hamilton, *Living Arts*, The Institute of Contemporary Arts, no. 2, 1963, 48–49. Courtesy of Richard Hamilton Estate.

1.12 following pages
Spread from "Persuading Image" article by Richard Hamilton, including the illustration of swimwear model Vikky Dougan's back. "Persuading Image" by Richard Hamilton, *Design*, February 1960, 30–31. Courtesy of University of Brighton Design Archives.

Symbols are interchangeable. Autos borrow
imagery from rocket missiles as readily as lighters
adopt the visual language of science fiction
or radios the terminology of sports cars.

The dead cert that came home last.

sales. If a design for industry does not sell in the quantities for which it was designed to be manufactured then it is not functioning properly.

The element in the American attitude to production which worries the European most is the cheerful acceptance of obsolescence; American society is committed to a rapid quest for mass mechanized luxury because this way of life
satisfies the needs of American industrial economy. By the early 'fifties it had
become clear in America that production was no problem. The difficulty lay in
consuming at the rate which suited production and this rate is not only high – **it
must accelerate**. The philosophy of obsolescence, involving as it does the
creation of short-term solutions, designs that do not last, has had its drawbacks
for the designer – the moralities of the craftsman just do not fit when the product's
greatest virtue is impermanence. But some designers have been able to see in
obsolescence a useful tool for raising living standards. George Nelson in his book
Problems of Design, states the case very forcibly: "Obsolescence as a process is
wealth-producing, not wasteful. It leads to constant renewal of the industrial
establishment at higher and higher levels, and it provides a way of getting a
maximum of good to a maximum of people". His conclusion is: "What we need
is more obsolescence not less". Mr Nelson's forward-looking attitude squarely
faced the fact that design must function in industry to assist rapid technological
development; we know that this can be done by designing for high production
rates of goods that will require to be renewed at frequent intervals.

The responsibility of maintaining the desire to consume, which alone permits
high production rates, is a heavy one and industry has been cross-checking. With
a view to the logical operation of design, American business utilized techniques
which were intended to secure the stability of its production. In the late 'forties
and early 'fifties an effort was made, through market research, to ensure that
sales expected of a given product would, in fact, be available to it. Months of
interrogation by an army of researchers formed the basis for the design of the
Edsel, a project which involved the largest investment of capital made by American business in post-war years. This was not prompted by a spirit of adventure –
rather it was an example of the extreme conservatism of American business at
the time. It was not looking to the designer for inspiration but to the public,
seeking for a composite image in the hope that this would mean pre-acceptance
in gratitude for wish-fulfilment. American business simply wanted the dead cert.
It came as something of a shock when the dead cert came home last. The *Edsel*
proved that it took so long to plan and produce an automobile that it was no
good asking the customer what he wanted – the customer was not the same
person by the time the car was available. Industry needed something more than
a promise of purchase – it needed an accurate prophecy about **purchasers
of the future**. Motivation research, by a deeper probe into the sub-conscious
of possible consumers, prepared itself to give the answer.

It had been realised that the dynamic of industrial production was creating an
equal dynamic in the consumer, for there is no ideal in design, no pre-determined
consumer, only a market in a constant state of flux. Every new product and
every new marketing technique affects the continually modifying situation. For
example, it has long been understood that the status aspect of car purchasing is
of fundamental importance to production. Maintaining status requires constant
renewal of the goods that bestow it. As *Industrial Design* has said: "post-war values
were **made manifest in chrome and steel**".[5] But the widespread re-

The current signs of sex, the outward and visible equipment of epic male and female, are the persuasive images of desirable glamour, creating for this era a formal background which profoundly influenced styling trends.

alisation of aspirations has meant that gratification through automobile ownership has become less effective. Other outlets, home ownership and the greater differentiation possible through furnishing and domestic appliances have taken on more significance. Company policy has to take many such factors into account. Decisions about the relationship of a company to society as a whole often do more to form the image than the creative talent of the individual designer. Each of the big manufacturers has a design staff capable of turning out hundreds of designs every year covering many possible solutions. Design is now a selective process, the goods that go into production being those that motivation research suggests the consumer will want.

Most of the major producers in America now find it necessary to employ a motivation staff and many employ outside consultants in addition. The design consultants of America have also had to comply with the trend to motivation research and *Industrial Design* reports[6] that most now have their own research staffs. This direction of design by consumer research has led many designers to complain of the limitation of their contribution. The designer cannot see himself just as a cog in the machine which **turns consumer motivations into form** – he feels that he is a creative artist. Aaron Fleischmann last year, in the same *Industrial Design* article, expressed these doubts: "In the final analysis, however, the designer has to fall back on his own creative insights in order to create products that work best for the consumer; for it is an axiom of professional experience that the consumer cannot design – he can only accept or reject". His attitude underrates the creative power of the yes/no decision. It pre-supposes the need to reserve the formative binary response to a single individual instead of a corporate society. But certainly it is worthwhile to consider the possibility that the individual and trained response may be the speediest and most efficient technique.

Design in the 'fifties has been dominated by consumer research. **A decade of mass psycho-analysis** has shown that, while society as a whole displays many of the symptoms of individual case histories, analysis of which makes it possible to make shrewd deductions about the response of large groups of people to an image, the researcher is no more capable of creating the image than the consumer. The mass arts, or pop arts, are not popular arts in the old sense of art arising from the masses. They stem from a professional group with a highly developed cultural sensibility. As in any art, the most valued products will be those which emerge from a strong personal conviction and these are often the products which succeed in a competitive market. During the last 10 years market and motivation research have been the most vital influence on leading industrialists' approach to design. They have gone to research for the answers rather than to the designer – his role, in this period, has been a submissive one, obscuring the creative contribution which he can best make. He has, of course, gained benefits from this research into the consumers' response to images – in package design particularly, techniques of perception study are of fundamental importance. But a more efficient collaboration between design and research is necessary. The most important function of motivation studies may be in aiding control of motivations – to use the discoveries of motivation research to promote acceptance of a product when the principles and sentiments have been developed by the designer. Industry needs greater **control of the consumer** – a capitalist society needs this as much as

5 *Industrial Design*, February 1959, page 79.
6 *Industrial Design*, February 1958, pages 34-43.

31

1.13
Photograph of Richard Hamilton (in the American football uniform) and his crew setting up for the photo shoot of the "Urbane Image" self-portrait used on the cover of *Living Arts*. The props included a 1963 Ford Thunderbird, a refrigerator stuffed with American food, a Wondergram, a vacuum cleaner, telephone, typewriter, and toaster all arranged on a background of high-gloss pink paper on the pavement outside the Taylor Woodrow Building, and photographed from an upper floor by Robert Freeman. Hamilton credited the photographer, the stylist, and himself as "producer" of the image. Courtesy of Richard Hamilton Estate.

"A THROW-AWAY ESTHETIC"

1.14
Photograph of Richard Hamilton and his crew
maneuvering a Mercury spacecraft capsule on loan
from Shepperton Film Studios into place for the
photo shoot of the "Urbane Image" self-portrait
used on the cover of *Living Arts*. Courtesy of
Richard Hamilton Estate.

"Stirring the pot of controversy"[83]

Considering how divergent Hamilton's article was from CoID's values and *Design*'s typical content, why did the magazine's editor commission the piece? In 1960 the magazine's editorship was in transition. Michael Farr, who had edited *Design* since 1952, became Chief Information Officer, and was in the process of handing the magazine's editorship to his deputy editor, John E. Blake, a young graduate of the Royal College of Art who had worked on *ARK*, the college publication. In February 1960 Blake had been editor of *Design* for only a couple of months. It is likely that, with the publication of Hamilton's piece, Blake was trying to define a new direction for the magazine and stake out the different terms of his editorship, while Farr was trying to generate new readers for the magazine, for which he now had a greater financial responsibility.[84]

Most historical accounts portray *Design* as merely a propagandist "mouthpiece" of the CoID, and there are certainly grounds for this view in the close parallels between CoID's values and the content of the magazine. Still, there was a tension at play in the pages of *Design* between the principles of CoID-endorsed good design on the one hand and the imperatives of critical journalism on the other. The editors were often under pressure from the Council's Information Division to include in the magazine CoID-approved consumer goods, especially those manufactured by its trustees, who were also advertisers and thus among CoID's sources of income.[85] By the early 1960s, advertising had assumed an increased influence on the magazine's editorial content. Particularly objectionable to the management were articles that seemed in any way "anti-British industry," and the fact that *Design*'s editorial was increasingly devoted to "overseas material," which presented "an almost hopeless task for gaining advertisements."[86] But editors like Farr and his mentee Blake had a journalistic appreciation for provocation, and saw their role as injecting lively debate into the journal's pages.

According to Ken Garland, who was *Design*'s art director at the time, Farr "relished stirring the pot of controversy."[87] Farr was interested enough in other design and consumer magazines to instruct Garland to prepare for him a monthly report on them. Like Hamilton, Garland appreciated the American magazines *Interiors* and *Industrial Design*. He recalled: "if I had a model it would be these magazines ... we used to receive these publications with great eagerness each month."[88]

Prior to joining *Design*, Farr had also worked as news editor for the *Architects' Journal* and the *Architectural Review*.[89] In *Architectural Review* during 1959 and 1960, critical debates were signaled with the use of yellow paper stock, red type, and attention-grabbing typographic devices such as starbursts, enlarged quote marks, and arrows, and led to opinionated letters printed in subsequent issues.[90] It is probable, therefore, that in asking writers like Hamilton, Banham, and Alloway to contribute to *Design* articles that challenged CoID's worldview, Farr and Blake were seeking to emulate *Architectural Review*'s debate-generating strategy in the hope of gaining more readers.

1.15
Cover of *Design*, The Council of Industrial
Design, September 1960, designed by Ken Garland.
Courtesy of Ken Garland and University of
Brighton Design Archives.

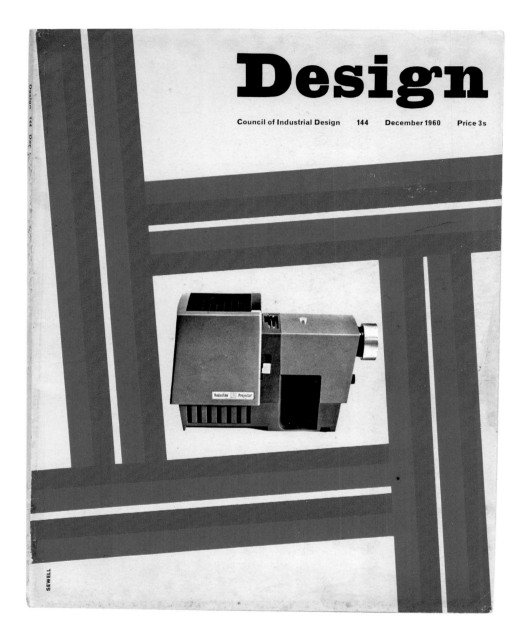

1.16
Cover of *Design*, The Council of Industrial Design,
December 1960, designed by John Sewell, art-
directed by Ken Garland. Courtesy of Ken Garland
and University of Brighton Design Archives.

1.17
Michael Farr, editor of *Design*, 1952–1960, and John
Blake, editor of *Design*, 1960–1967, photographed
by John Garner, *Design*, January 1960, 27. Courtesy
of University of Brighton Design Archives.

The art direction of "Persuading Image"

Within the pages of the February 1960 issue of *Design*, Hamilton's article was emphatically flagged. It was the lead article. Its first page was printed on bright yellow paper, and color tints were used, at a time when color tended to appear in the magazine only when manufacturers paid for it in order to better show off their products.[91] Garland recalls that the bolding of key phrases, such as "control of the consumer" and "plush at popular prices," was to "enliven and emphasize" the text, and was done in consultation with Hamilton.[92] Hamilton's use of imagery was also unique within the pages of *Design*. The images he and Garland selected to illustrate his piece floated alongside the text allusively rather than as directly referenced examples. Also hovering were Hamilton's enigmatic captions, which quoted advertising slogans and editorial hyperbole as a form of poetry. Beneath a selection of images of car detailing and a page excerpted from *Look* magazine, for example, is the text: "'Functionalism is not enough for Americans,' says the page from *Look*, and the automobile body designer knows it. High fashion stylists in metal use the symbols of speed, sex and status to gain sales appeal."[93] This kind of unfiltered sampling of American advertising and editorial language sits uneasily in the pages of *Design*, but found a more fitting home in Hamilton's artworks and more creative pieces of writing (such as his "Urbane Image" article published in *Living Arts* in 1963), where ambiguity lends the works and prose their tension.

The contentious nature of Hamilton's article was suggested in Blake's editorial introduction to the issue. Under the headline "Consumers in Danger," Blake primed his readers by promising them a "controversial" article with a conclusion "of a form of economic totalitarianism not greatly dissimilar from Orwell's terrifying prophecy." Blake summoned the force of George Orwell's novel about totalitarian ideology, *Nineteen Eighty-Four*, which had been published in 1949 and had sustained the public's attention, as its narrative seemed to be confirmed by actual events of the Cold War. He concluded his introduction with the rhetorically loaded question: "Do we believe it is more important for industry, and the designer, to serve the real needs of the consumer, or are we content with the prospect of the consumer becoming a pawn in the grip of an economic master who rules exclusively to serve his own ends?"[94]

In his editorial leader, Blake drew readers' attention to what he perceived to be the key phrase of the piece: "design a consumer to the product." In an effort to extend the article's lifespan across several issues, just as the *Architectural Review* was doing so successfully with its Criticism features, Blake promised a continuation of the debate in a subsequent issue, where he planned to publish the "comments of designers and design critics from Europe and America on the issues raised in this controversial article."[95]

The readers respond

Design magazine invited responses to Hamilton's article from a select group of American and European designers, manufacturers, and critics, which appeared in the magazine's

June issue. Most objected to what they perceived to be Hamilton's lack of social responsibility and his complicity with reviled American values. Industrial designer Misha Black said: "The designer can admittedly 'maintain a respect for the job and himself while satisfying a mass audience,' but only while he retains some respect for the civilization of which he is a part; if he ceases to be concerned with real values in society then he becomes a polite equivalent of the dope peddler who also satisfies a social need."[96] D. W. Morphy, of British home appliances firm Morphy Richards, considered the design Hamilton talked about "false design," and hardly likely to deceive the public.[97] Alberto Rosselli, editor of the Italian magazine *Stile Industria*, was reported as saying: "The Hamilton prescription is immoral in that it might lead to indiscriminate use of persuasion."[98] Even George Nelson, whose articles in *Industrial Design* had inspired Hamilton, commented that Hamilton's conclusions were "depressing, even nauseous."[99]

The discussion continued in the form of readers' letters in subsequent issues, but Hamilton himself was given the last word. He wrote: "The phrase that caused the alarm, 'designing the consumer to the product,' is a redefinition of a well-known process; for the ultimate political evil it was called fascism, when directed at purely commercial objectives it is called salesmanship, without the moral overtones it is known as education. We are all concerned, in one way or other, with the conversion of others to a point of view."[100] Hamilton was particularly keen to have the last word with regard to the comments submitted by Banham, then assistant executive editor of the *Architectural Review* and Hamilton's intellectual sparring partner in the Independent Group. Banham was dismissive of Hamilton's arguments, and seemed to be defending his own preserve of writing about industrial design. He suggested that one benefit of obsolescence could be the creation of a situation in which "fine art designers who believe a 'good design is forever,' will decide that product design is beneath their contempt, and get out, leaving the field to men far better qualified to realise the satisfaction of consumer wants with a far clearer sense of the product designer's position as the *servant* of his mass public."[101] Hamilton responded to his "critics" generally except in the case of Banham, whom he singled out for direct rebuttal: "[Banham's] reading was as slipshod as any since he repeats much of what I said in a tone of contradiction, but he is so much a democrat that he equates 'controlled' with 'being pushed around.' If his conception of democracy is carried much further there is a danger of his becoming conservative."[102]

1.18 following pages
Opening spread from "Persuading Image"
article by Richard Hamilton, *Design*, February
1960, 28–29. Courtesy of Ken Garland and
University of Brighton Design Archives.

This article is a shortened version of The Design Image of the 'Fifties, *a lecture given by the author at the Institute of Contemporary Arts recently. Richard Hamilton used illustrations from the pages of American magazines to show how an image of the "fabulous 'fifties" was being created by American designers, advertisers and industrialists to instil in the consumer a "desire for possession". The craving to consume must be fostered to meet the needs of America's industrial economy, but it means that the designer must adjust himself to a new set of values if he is still to play a creative role. Market research aims to tell the industrialist what the consumer wants, but it cannot forecast what the public is likely to want in four or five years' time. The designer in the 'sixties, Mr Hamilton suggests, will be working with industrialists and motivation researchers to "design a consumer to the product". Not every designer will feel, to quote Mr Hamilton, that "within this framework the designer can maintain a respect for the job" and find ". . . more precise solutions to the needs of society". In a subsequent issue this year* DESIGN *hopes to publish the comments of designers and design critics from Europe and America on the issues raised in this controversial article.*

PERSUADING IMAGE

RICHARD HAMILTON

The 'fifties have seen many changes in the human situation ; not least among them are the new attitudes towards those commodities which effect most directly the individual way of life – consumer goods. It is now accepted that saucepans, refrigerators, cars, vacuum cleaners, suitcases, radios, washing machines – all the paraphernalia of mid-century existence – should be designed by **a specialist in the look of things**. Of course, the high power virtuoso industrial designer is not a new phenomenon – Raymond Loewy and Walter Dorwin Teague have been at it for a good many years. William Morris and Walter Gropius realized the potential. What is new is the increased number of exponents, their power and influence upon our economic and cultural life. Design is established and training for the profession is widespread.

The student designer is taught to respect his job, to be interested in the form of the object for its own sake as a solution to given engineering and design problems – but he must soon learn that in the wider context of an industrial economy this is a reversal of the real values of present-day society. Arthur Drexler has said

28

"It's a rich man's kind of car...but no kind of price!"

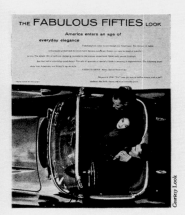

THE **FABULOUS FIFTIES** LOOK

America enters an age of
everyday elegance

Courtesy Look

Courtesy Look

"Functionalism is not enough for Americans", says the
page from *Look*, and the automobile body designer knows it.
High fashion stylists in metal use the symbols of speed, sex
and status to gain sales appeal.

of the automobile "Not only is its appearance and its usefulness unimportant . . .
What is important is to sustain production and consumption". The conclusion
that he draws from this is that "if an industrialised economy values the process by
which things are made more than it values the thing, the designer ought to have
the training and inclinations of a psycho-analyst. Failing this he ought, at least, to
have the instincts of a reporter, or, more useful, of an editor."[1]

The image of the 'fifties shown here is the image familiar to readers of the
glossy magazines – "America entering the age of everyday elegance"; the image
of *Life* and *Look*, *Esquire* and the *New Yorker*; the image of the 'fifties as it was
known and moulded by the most successful editors and publicists of the era, and
the ad-men who sustained them – "the fabulous 'fifties" as *Look* has named them.
Being **"plush at popular prices"** is a prerogative that awaits us all. Whether we like it or not, the designed image of our present society is being realized
now in the pages of the American glossies by people who can do it best – those
who have the skill and imagination to create the image that sells and the wit to
respond humanly to their own achievements.

The present situation has not arrived without some pangs of conscience. Many
designers have fought against the values which are the only ones that seem to
work to the economic good of the American population. There is still a hangover
from the fortyish regret that things do not measure up to the aesthetic standards
of pure design ; the kind of attitude expressed in 1947 by George Nelson when he
wrote : "I marvel at the extent of the knowledge needed to design, say, the *Buick* or
the new *Hudson* – but I am also struck by my inability to get the slightest pleasure
out of the result".[2] There has since been a change of heart on both sides ; on the
part of the designers, the men who establish the visual criteria, towards a new
respect for the ability of big business to raise living standards – and an appreciation, by big business, of the part that design has to play in sales promotion. What
was new and unique about the 'fifties was a willingness to accept a new situation
and to **custom build the standards** for it.

There is not, of course, a general acceptance of this point of view. Some designers, especially on this side of the Atlantic, hold on to their old values and are
prepared to walk backwards to do so. Misha Black goes so far as to suggest that
advanced design is incompatible with quantity production when he says: "If
the designer's inclination is to produce forward-looking designs, ahead of their
acceptability by large numbers of people, then he must be content to work for
those manufacturers whose economic production quantities are relatively small"[3].
While Professor Black was consoling the rearguard for being too advanced Lawrence Alloway was stressing the fact that "Every person who works for the public
in a creative manner is face to face with the problem of a mass society"[4]. It is just
this **coming to terms with a mass society** which has been the aim and
the achievement of industrial design in America. The task of orientation towards
a mass society required a rethink of what was, so convincingly, an ideal formula.
Function is a rational yardstick and when it was realized in the 'twenties that all
designed objects could be measured by it, everyone felt not only artistic but right
and good. The trouble is that consumer goods function in many ways ; looked at
from the point of view of the business man, design has one function – to increase

1 Foreword to *Problems of Design*, George Nelson, Whitney Publications; Inc, Alec Tiranti Ltd, £3 12s 6d
2 A lecture at The Chicago Institute of Design, reprinted in *Problems of Design*.
3 *The Honest Designer*, the Percy Wells Memorial Lecture given by Misha Black to the Technical College for the
Furnishing Trades, Shoreditch, 1958.
4 *Architectural Design*, February 1958, page 84.

29

Writing about the effects of Hamilton's "Persuading Image" article, design historian John Hewitt concludes: "To a Council that had always preferred the idea of serving a consumer, of responding to his/her needs, not those created through market research or advertising, the very idea of controlling the consumer and of integrating him/her totally into the market processes in order to meet different, purely commercial objectives, was total anathema."[103]

And yet, it had also been the CoID's longstanding mission to "mould" British consumers by seeking to educate them in the principles of good design. *Design* magazine, specifically, under pressure to sell more issues, was beginning to show curiosity about, if not exactly to "depth probe," its own consumers. While Hamilton's article was being published, *Design*'s managers were in the process of employing a market research firm to conduct reader surveys. Mass Observation's 1961 report on *Design* magazine's readership unearthed a litany of grumbles about the magazine's form and content.[104]

It turned out that the consumer of *Design* was harder to shape than its managers thought. Readers had specific views about *Design* and how it could be improved. Seventeen percent thought it should contain more about readers' own jobs. It was deemed by some "too arty and academic"; or "not sufficiently up-to-date"; while others objected to the criteria it used to judge good design. Others thought the magazine "should have more expert reports"; that it was "badly written"; "needs more outside writers"; and "should be aimed more at the man in the street."[105] "My main criticism," said a design consultant, homing in on a growing public perception of the CoID as elitist and out of step with the times, "is that it is too snooty about everything. There is no link made—or no effort at a link—between the designer and the ordinary people. It fails because it relies too much on snob value. I feel that a 'Design Establishment' is emerging which is far too tight."[106]

One of the subscribers interviewed by Mass Observation, a production planning manager in an engineering firm, objected to the "parochial tendencies" of the magazine, and said he wanted "more variety among contributors."[107] In explaining his request he unconsciously quoted Hamilton's "Persuading Image" article, saying: "[*Design*] should deal with the social side of design. You can now design the customer to the product as well as the product to the customer."[108] It appeared that however heterodox Hamilton's argument might have been in the context of CoID's anti-commercial, socially and morally driven view of design, among *Design*'s actual readership its message had hit home.[109]

Hamilton did not consider himself a design critic. "I've always thought of myself as a design hobbyist," he later said.[110] And yet he had clear views about his role as an educator, which do seem to have translated into his critical writing: "It is for us as teachers to promote in the youth we teach a healthy suspicion of all dogma, whether it is politically oriented or aimed at fixing the pattern of our culture," he said in a 1960 lecture to the National Union of Teachers.[111] He believed that in order to achieve "freedom of choice ... the youth of today" should be made fully aware of the techniques of mass media, "whose products

"A THROW-AWAY ESTHETIC"

they already know and appreciate."[112] Hamilton's perceptive analysis of the techniques of image making, the social and economic implications of mass-produced goods, and the inevitability of expendability helped to challenge the main current of design discourse in postwar Britain with a level of authority seldom found among the writings of establishment-sanctioned design commentators of the period.

PART TWO: THE APPLIED LIFE OF PRODUCTS IN *INDUSTRIAL DESIGN* MAGAZINE, US, 1955–1960

Industrial Design, "the most professional of design magazines"[113]

While Hamilton's article ran counter to the ethos of the British CoID and the content of its house magazine, it connected quite closely to the kinds of preoccupations of American design discourse being exercised in the New York-based independent trade publication *Industrial Design*. This magazine was among the sources of images and articles that members of the Independent Group used for their lectures, articles, and artworks—and was referenced in particular by Banham, who dubbed it "the most professional of design magazines."[114] Under the coeditorship of Jane Thompson and Deborah Allen, *Industrial Design* offered a pluralist view of product design that acknowledged the existence of "a mass culture, in which artifacts are produced under completely new circumstances," and the reality that "we have in mass-produced objects a new kind of folk art in a new dimension: an anonymous, or group-oriented expression of the twentieth century in terms of practical needs—which is not by all the people, but at least for the people."[115]

Industrial Design had begun life in 1941 as a column in *Interiors*, a magazine devoted to the interior design profession. Upon the advice of designer George Nelson, its publisher Charles E. Whitney decided to develop the column into a publication aimed at industrial designers "concerned with product planning, design, development and marketing."[116] In 1954 the "Industrial Design" column editors Jane Thompson (then Fiske Mitarachi) and Deborah Allen became the new magazine's first editors, with Nelson as editorial contributor and advisor. Nelson's design office was in the same building, and he seems to have had some considerable influence on the content and ethos of the magazine. Thompson remembered: "He would decide what he wanted to write and once in a while he decided what you wanted to write."[117] Nelson, primarily a designer and at the time design director for Herman Miller Furniture Company, also had experience as a writer and editor. He was co-managing editor of *Architectural Forum*, a contributor to both *Fortune* and *Interiors*, and in 1959 his collected essays were published by Whitney in the book *Problems of Design.* When Ralph Caplan joined the magazine as an associate editor in 1957, he tutored himself on design and its issues by reading this book, copies of which were stored in a crate by the men's room.

1.19 above and following pages
Covers of *Industrial Design*, October 1954, and August
1955, February 1956, designed by Martin Rosenzweig.
Courtesy of F&W.

INDUSTRIAL DESIGN

4

August 1955   two dollars a copy

Fasteners and fastening techniques   Pointers for designing sheet-formed products.

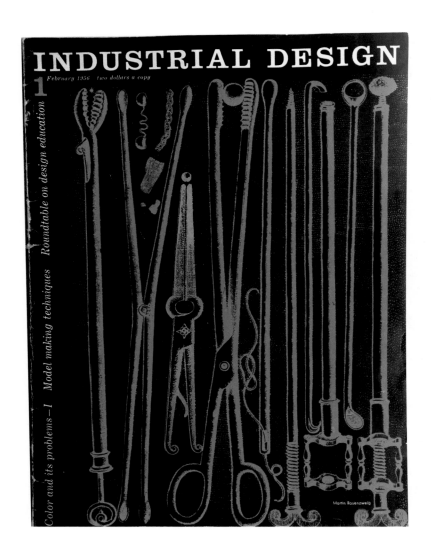

INDUSTRIAL DESIGN

February 1956   two dollars a copy

1

Roundtable on design education

Model making techniques

Color and its problems—I

Martin Rosenzweig

"A THROW-AWAY ESTHETIC"

In his "Publisher's Postscript" to *Industrial Design*'s first issue in February 1954, Whitney explained his perspective on the genesis of the bimonthly journal: "The establishment of a new magazine was made almost mandatory by a series of developments in the last decade—the ascent of the product designer to a position of executive authority in industry; the vigorous demand by designers for a publication edited exclusively for them; and more particularly, the enlightening contacts we made at Walter P. Paepcke's Aspen Design Conference two years ago."[118]

The magazine went on to develop a close relationship with the International Design Conference at Aspen (IDCA) through reporting its activities, republishing its papers, and the magazine editors' involvement as moderators and conference board members. During these years both the conference and the magazine campaigned for greater recognition of design's value to business and society, and sought to promote the significance of design "as a unique, autonomous function in the overall industrial operation—on parity with engineering, manufacturing and sales."[119]

Nelson's own role as design director of Herman Miller was an exemplar of this mission to elevate the standing of the designer in the "industrial operation." In the first issue of *Industrial Design*, Nelson wrote: "The designer functions as a member of the top policy group and his recommendations carry the same weight as those of the production and sales executives."[120]

In addition to its close ties to the IDCA, *Industrial Design* operated within a network of other contemporaneous general interest magazines such as *Harper's*, *Collier's*, *House & Home*, and the *New Yorker*, international design magazines like *Design* in Britain and *Domus* in Italy, and museums, especially the Museum of Modern Art. *Industrial Design* frequently commissioned writers and republished articles from other magazines, and from recently published design books from the Whitney publishing stable, while its editors participated in, and reported on, debates on styling and "good design" at MoMA. Despite such exchanges, the particular quality of *Industrial Design*'s engagement with its subject matter was unique within this network. A 1958 panel, organized to discuss an exhibition of "Twentieth Century Design from the Museum Collection," moderated by *Industrial Design*'s then-consulting editor Jane Thompson and recorded in the magazine under the title "Design as Commentary," revealed some of the differences between the Museum's and the magazine's conception of design. Arthur Drexler, director of MoMA's Department of Architecture and Design since 1956, was one of the panelists, and stated that the museum's collection purposefully excluded "those mass-produced objects supposed to be characteristic of our high standard of living. There are no television sets, no refrigerators, no telephones, and only a few mechanical appliances—not because such objects are intrinsically unworthy but rather because their design seldom rises above the vulgarity of today's high-pressure salesmanship."[121] *Industrial Design*, on the other hand, devoted a whole section each month to analysis of such domestic appliances. Drexler went on to observe: "The Museum's collection is not concerned with persuading people to use objects, to buy them, to consume. Our interests are concerned primarily with art."[122] While *Industrial Design* certainly promoted design, its editors also critiqued it. They considered formal beauty too limiting a criterion, however, and focused instead upon the way products worked, how they were used, and what they said about "a heavily goods-oriented society."[123]

The US did not have a government agency like the British CoID to campaign for the importance of design to industry, and so this job was left to entrepreneurial individuals who had been instrumental in the formation of the country's industrial design profession. In the 1940s they had started to assemble into professional organizations. The Society of Industrial Designers had been established in New York in 1944, initially with fourteen members, including Raymond Loewy, Norman Bel Geddes, Walter Dorwin Teague, Henry Dreyfuss, Donald Deskey, Harold Van Doren, and Russel Wright.[124] The group initiated an annual awards scheme and produced an occasional annual publication called *US Industrial Design*, but they did not possess the journalistic drive to create news stories, nor the distance from the profession necessary to a critical stance. Thompson and Allen, at the helm of *Industrial Design*, on the other hand, helped to pioneer a distinctively American, mass-market product design criticism, fueled by their personal beliefs, intellectual backgrounds, and experiences as both professional working women and homemakers.

Televisions, refrigerators, and telephones

The interrelated philosophies of relativism and pragmatism permeated much of liberal intellectual American culture in the postwar period. In 1950 the historian Henry Steele Commager praised pragmatism, describing it as deriving directly from the country's historical experience and becoming, in the twentieth century, "almost the official philosophy of America."[125] The sociologist Daniel Bell recommended an eschewal of utopian ideologies that had been tainted by totalitarianism, and adherence, instead, to a quintessentially American tradition of sober, prudent practicality; while historian Daniel J. Boorstin advocated for a "doctrinally naked," and therefore flexible, America able to accept "the givenness of experience."[126] Disturbed by the activities of anticommunist ideologists in the 1940s such as Senator Joseph McCarthy, the American intellectual critical community, typified by such groups as the New York Intellectuals (who included essayists such as Lionel Trilling, Harold Rosenberg, and Daniel Bell), abandoned what Neil Jumonville has termed "their earlier ideological and faith and prophetic partisanship," and adopted "a more modest and precise outlook based on reason, analysis, and pragmatism."[127]

Thompson and Allen, while not overtly political, deployed a similarly rationalist, pluralist, and nonpartisan outlook to the New York Intellectuals. But where the latter saw mass culture as threatening their professional status, Thompson and Allen were not afraid to deal very directly with it. They viewed designed products and their impact on everyday life as ripe territory for literary exploration. Throughout the pages of *Industrial Design* their version of pragmatic relativism is evident in their frequent use of personal experience to illuminate the specifics of a product, their innovative use of explanatory diagrams and "how-to" guides, and their refusal of aesthetic absolutes and prevailing ideologies such as "good design."

Thompson wrote several articles about her approach to evaluating industrial design. She dismissed the use of set standards, which she termed "automatic evaluation": "The end result is a code-book of styles; no one need bother to think for himself as long as he has the rules firmly memorized."[128] Her preferred method was "creative evaluation," which necessitated an immersive understanding in order to "look at a thing and understand not how it conforms to existing rules, but what new rules it may be suggesting for the future."[129] Thompson believed that taste was a "smokescreen" that prevented one getting to the "deeper implications" of design, a "substitute for evaluation, rather than a basis of evaluation." In a July 1957 editorial preface to *Industrial Design*, she further expanded her relativist position on assessing design: "[Design] can be judged 'superior' or 'inferior' only on its own terms. I am aware that moralists do not enjoy this point of view. It is hard not to rely on the crutch of our own absolute Good and absolute Bad. Yet if one is serious about judging design, the task, as in viewing all art, is to overcome the temptation to judge its subject matter alone, or its moral value, and to sense its vigor, its aptness, its communication."[130]

1.20
Jane Thompson, photographed by Richard Pousette Dart, and used to illustrate "Working in a Man's World," by Jane Fiske McCullough, *Charm*, November 1957, 86. Courtesy of Jane Thompson.

Jane Thompson grew up in Larchmont, Westchester County, New York, the daughter of an air-conditioning and refrigeration engineer, who also edited a trade association magazine and "wrote a lot." Reflecting on the role of writing in her childhood years, she said: "I was used to the idea of sitting at a typewriter and grinding things out."[131] Thompson studied at Vassar College, a prestigious women's liberal arts college in Poughkeepsie, New York, and pursued graduate studies at New York University's Institute of Fine Arts. She began her career as secretary to the architect Philip Johnson, who was then the curator and head of MoMA's Department of Architecture and Design. She soon transitioned to the role of acting assistant curator. Of this period, during which the museum staged the first US Mies van der Rohe exhibition (1947) and installed the Marcel Breuer House in its garden (1949), she has observed: "It was an education in the history of architecture and its future, and it also helped me to develop my critical sense."[132] In 1949 Johnson hired Arthur Drexler, architecture editor from *Interiors* magazine, to be a curator, and Thompson took his vacated position at the magazine.

1.21
Photograph of Deborah Allen's Westport family home, by Oliver Allen, c. 1948. Courtesy of Oliver Allen.

"A THROW-AWAY ESTHETIC"

Deborah Allen was an associate editor at *Interiors*, and Thompson identified her as a likely collaborator. Like Thompson, Allen had grown up around writing. She believed her interest in design, her opinionated nature, her taste and her work ethic derived from a cultured family upbringing and some interesting female role models. Her aunt was Ethel B. Power, editor of the home decorating magazine *House Beautiful* between 1923 and 1934, and her aunt's partner was the architect Eleanor Raymond. Allen's mother, Dwight Hutchinson, worked as a copywriter at J. Walter Thompson, and then as a freelance writer for women's magazines. Allen's childhood home in Boston was filled with books and magazines about design and interiors, and designed objects her mother had brought back from trips to Sweden. While studying art history at Smith College, a liberal arts college for women, Allen wrote for the college newspaper, and the writer Mary Ellen Chase, who was in residence at Smith at the time, read her work and sought her out. "She said, 'don't do anything that will teach you to be glib. Take your writing seriously.' I liked that," Allen recalled.[133] After graduating, Allen worked for a short while at the Metropolitan Museum of Art and married Oliver Allen, an editor at *Life* and son of Frederick Lewis Allen, editor of *Harper's*.

In 1953, when *Interiors* publisher Whitney asked Thompson to edit a new magazine for industrial designers, she asked that Allen be her coeditor. The women were given a small budget and the sole mandate that the magazine should be as graphically bold and handsome as *Fortune* magazine was at the time. "He wanted flashy gate folds," Thompson recalled.[134]

Unfettered by any institutional or prescriptive viewpoint on design, Thompson and Allen set out to build from scratch a magazine for the industrial design profession informed by their own educational backgrounds in the humanities, their professional experience as journalists, and their domestic responsibilities as wives and mothers. (These were not insignificant—Thompson married four times and had two children and seven stepchildren, while Allen had five children.)

The magazine's business model was based on a mixture of advertising and subscriptions, which rose from 5,910 in 1955 to around 10,000 by 1959. Advertisements (mainly for materials producers and fabrication services such as Arabol Adhesives, Marco Polyester Resins, Chicopee Specialty Weaves, Aluminum Extrusions, and Dupont, and a handful of furniture companies like Knoll) were mostly grouped in the front-of-book, with the editorial preface marking the start of the feature well.[135] As Ralph Caplan, who became editor of the magazine in 1958, remembers it, the publisher was disappointed with the advertising revenue; he had mistakenly thought that industrial designers specified furniture and materials, just as interior designers did, and that he could sell advertising on the same basis as he did at his other magazine, *Interiors*. Caplan observed: "Although an industrial designer might specify that a product be made of aluminum, he was not empowered to choose Reynolds or Alcoa."[136]

1.22
Jane Thompson, editor, and Charles Whitney,
publisher, in the *Industrial Design* office, c. 1957.
Courtesy of Jane Thompson.

While the stated purpose of *Industrial Design* was to elevate the standing of the designer in the realm of commerce, for Allen and Thompson there was another goal, expressed through their chosen subject matter and examples, and that was "to connect designers to consumers and users—the applied life of the product."[137] Unlike *Design* magazine in the UK, however, where the consumer was conceived of as rational and willing to be educated by the editors' superior taste and knowledge, Allen and Thompson wrote for a consumer who also had irrational whims and emotional concerns. Thompson recalled: "we perceived things that we needed that were not being answered by the designer. We saw from a consumer's perspective the way a product works or doesn't work, or pleases or offends."[138]

In an editorial about taxi design, for example, they described an industrial designer in their own terms, thus subtly guiding their readership toward a similar view: "He's not so much a stylist—a man who slaps jumbo grilles and speedlines on another fellow's chassis—as a skilled and critical taxi rider, professionally fitted to give a roadworthy chassis a body worthy of human occupation."[139] Their choice of the terms "taxi rider" and "human occupation" here were key to their own guiding principles as critics: designers should be physically familiar with the use of the things they are designing, and concerned for the bodily and emotional well-being of other users.

1.23
Jane Thompson, Deborah Allen, and Allen's son, photographed by Oliver Allen, near the East River, New York, c. 1952. Courtesy of Oliver Allen.

The articles Thompson and Allen commissioned addressed a wide range of subjects, from bathrooms and plastics to tractors and design planning, and were characterized by deep research, clear exposition of complex technical issues, and extensive annotation. In addition to the staple fare of a design magazine, such as product reviews and issue-based features, Allen and Thompson introduced a wide array of unfamiliar article formats, including historical surveys of product types, cartoon interludes, photographic portfolios, book extracts, profiles of designers, and elaborate graphic devices such as timelines and charts. Allen had initiated such approaches while still at *Interiors* magazine. For her account of a 1950 MoMA panel discussion about the aesthetics of car design, she integrated condensed extracts of the panelists' arguments (not omitting their jokey quips) with images of the cars being discussed and diagrams of their components, adorned with pointing hand symbols and hand-drawn arrows. Her piece conveyed the dynamic nature of a live conversation and the voiced opinions of the participants far more directly than a linear report.[140] Allen continued to develop her visual article formats at *Industrial Design*. The article "What's So Special about Plastics?," for example, was laid out as a series of extended picture captions on spreads edged with binder file markings, suggesting its practical use in the design studio. In 1958 the designer Walter Dorwin Teague wrote in to congratulate the magazine for an article titled "Is This Change Necessary?" by Richard Latham, indicating one of the ways the magazine was used in a design studio: "I have asked all our partners here to read Latham's article—exceptionally well written by the way—and I shall read it again myself and keep it at hand for ready reference."[141]

Other readers' letters commended the magazine's range of formats. Raymond Loewy, probably the best-known designer in the US at the time, applauded the editors for "the variety of methods you are employing to report design activities—as projects, as individual case histories, as analyses of an office's operating techniques, and as aesthetic critiques."[142]

Thompson and Allen sought to explain complex ideas and technical processes through visual storytelling. The narrative of an article often continued into the image captions; manufacturing processes were broken down into digestible steps illustrated with cartoons; photographs of cars were silhouetted, cropped to highlight features and grouped for comparison. Of the other design magazines of the period, Thompson recalled: "There was no sense of energy, no attempt to convey ideas through the way you place things on a page, or how you use the type."[143] Thompson and Allen were unhappy with the art director of the first few issues, the acclaimed graphic designer Alvin Lustig, complaining that he was "too stiff" and resistant to a conception of page layouts as news-driven and visually animated compositions. "We wanted scale, changes of scale, big type, and a newsiness," said Thompson.[144]

To identify contributors, the editors used portrait photographs and short, familiarly written biographies. Nelson's design consultancy was described as having "an uncheckable tendency towards expansion," and contributors John W. Freeman and Alexandre

Georges were characterized as "looking as apprehensive as a couple of dicks." Such language signaled the editors' informal authority—their insider knowledge of their contributors beyond the bland facts of their official résumés. In the first issue, a series of cartoons by the illustrator Robert Osborn and Thomas B. Hess, editor of *Art News* and an exponent of biography-based criticism, satirized the stereotypes and pretensions of such résumés in portraits and fake biographies of Will C. Werk, Asa U Waite, Cozz McFields, and Rram de 'Vhwh (To pronounce 'Vhwh, say 'leave whey,' then omit 'lea' and 'hey.')"[145]

"Dear Sirs": the significance of gender

Despite their authority and desire to break new ground, the fact remained that the magazine catered to an almost wholly male readership of designers, engineers, and executives. Gender inequality was rife in the design industry and in society at large in the 1950s. As a result, letters to the editors were addressed "Dear Sirs"; the magazine's female writers were rarely mentioned in the list of contributors; not a single woman designer was profiled in at least the first decade of the magazine; and in 1957 Thompson noted that 80 percent of her appointments and interviews in the previous six months had been with men.[146] One of the most pronounced examples of the gender divide, against which Thompson and Allen's own careers appeared in stark relief, was in a report of the American Society of Industrial Designers' fourteenth annual conference, which "ended with a luncheon panel of designers' wives, each with her own idea of how and why to be one."[147]

Thompson and Allen brought a feminine perspective to bear on their subject matter—not in a politicized manner, but through what Thompson termed "an experienced and educated female instinct."[148] She said: "Women can look at a sharp object and know immediately that someone will get hurt with it. Men will never see it that way."[149] This maternal sense of danger was a recurring trope in the pages of *Industrial Design*. In her car reviews Allen would point out "the sharp edge" of the overhanging cowl of a Buick, which "looks as dangerous as the knobs it is supposed to shield," or car ashtrays which when opened make the dashboard turn menacing, since they are "frequently jagged edged and sticky."[150]

Thompson and Allen brought to traditionally masculine subject matter, such as cars, power tools, tractors, DIY, and plumbing, a point of view based on their domestic experience. And they brought that domestic experience, direct from their own homes and those of their friends, as subject matter into the pages of the magazine. The idea of changing lifestyles in the home, for example, became the focus of articles. "We knew that the separation between the dining room and the kitchen was breaking down," said Thompson, and to demonstrate this they staged a photograph at some architect friends' apartment in Greenwich Village, showing the family eating a meal in the kitchen.[151]

Thompson believed that she and Allen managed to "turn the female perspective to natural advantage in interpreting design. Our articles were informed not only with hard facts and real news, but also with the insights and attitudes of designers' ultimate customers—the female purchasers and users of products. This editorial pluralism built a perspective that no other design publication could offer to this special audience."[152] In an article titled "Working in a Man's World," which she wrote for *Charm* magazine in 1957, Thompson tried to convince working women that their female characteristics—"instinctual nurturing qualities," attention to detail, and insights from humble daily experience—were actually assets in the businesses where they worked. While such advice may seem conservative when considered in the light of burgeoning second-wave feminism, fueled by the 1963 publication of Betty Friedan's *The Feminine Mystique*, Thompson's own career and those of her female colleagues provided robust exemplars. Thompson stayed at the magazine as editor in chief until 1959, and as consultant editor until 1964. She went on to become a director of the Kaufmann International Design Awards, develop research on the history of the Bauhaus, join the board of directors of the International Design Conference at Aspen, and chair three of its conferences. She switched her focus to urbanism when she married the architect Ben Thompson and collaborated with him on many projects including the concept planning of the 1976 renovation of Boston's Faneuil Hall Marketplace, the running of the restaurant Harvest, and the influential store Design Research. Among the female writers for the magazine, Ann Ferebee went on to found and direct the Institute for Urban Design, and Ada Louise Huxtable became the first architecture critic for the *New York Times* in 1963. Even within such careers, the spheres of home and work were not separate, but inextricably entwined.[153]

Thompson and Allen co-wrote much of the magazine's copy, especially the editorial prefaces, and enjoyed a symbiotic working relationship. Allen's husband worked on weekends, closing the book at *Life* magazine on Saturday nights. Allen had to stay home to look after the children, so the women would work at her apartment on Beekman Place. They wrote articles collaboratively—rather like playing a game of hangman, Thompson recalled. Thompson would write one line and Allen the next, using an Olivetti typewriter. "And we'd write all the way through until we got something and then probably one person would patch it up, and then the other person would read it and patch it up some more. Our thinking was always in parallel and going in the same direction."[154]

"The editorial effort itself is a critical one"[155]

Allen and Thompson conceived of the entire project of editing the magazine as a form of criticism. Thompson devoted her April 1957 editorial preface to that topic, provoked by a reader who had written in to say: "It is not the business of the magazine to act as critic."[156]

Thompson and Allen believed that self-knowledge, which demands hard work, was essential to navigating the contemporary American consumer landscape and to outwitting "would-be manipulators." In her review of Vance Packard's book *The Hidden Persuaders*, Thompson wrote:

"A THROW-AWAY ESTHETIC"

Now there is no denying that Americans today are living out their lives, and their needs, through material symbols: the fins and portholes serve a deep-seated purpose in leading consumers into new social realms—imagined or real. But [Packard] reserves not one word of comment for the irrational consumer, and the ambitions and insecurities that drive him into the arms of businessmen. Is the condition the fault of merchandisers? Or are the merchandisers, rather, a symptom that people themselves might do well to examine?[157]

Thompson and Allen were also attentive to the needs of consumers of criticism, who included designers. They suggested that a designer had need of critics in order to develop his own critical faculties: "It is here that a magazine edited for him—continually studying his work and his problems—can be of some service. By expressing considered opinions and evaluating our motives for having them, the editors of Industrial Design hope to offer not only the news that each reader needs, but one set of views to help him form his opinions and examine *his* motives for doing what he does."[158]

Reflecting on this impulse in later in her life, Thompson said: "I think critical writing … is about trying to explain something so that the other person could have an opinion or evaluate it as well as you."[159] Thompson, Allen, and other writers, like the British critic Banham, did this by making their critical process accessible and visible, often taking readers through it with them step by step, with the intention of empowering readers to critique design for themselves.

Deborah Allen's lush, situational car reviews

By the mid-1950s, the American automobile industry, based in the Midwestern city of Detroit, had reached a plateau in technological developments to offer consumers. In order to compete for market share, the major companies, Ford, General Motors, and Chrysler (or "the Big Three," as they were called), put their resources into applying styling to the body shell of the car, focusing on details such as grilles, lights, fenders, tail fins, and chrome trim and painted metal strips, and into marketing these incremental style changes in their new models, using the women's fashion industry as inspiration. By 1957, General Motors was offering seventy-five body styles in 450 trim combinations.[160] Toward the end of the decade the automakers were bringing out new body shells every year, and these excesses were attracting criticism from all quarters.[161]

In articles such as "The Safe Car You Can't Buy," published in the Nation in 1959, Ralph Nader drew attention to the safety concerns and inconveniences (such as their inability to fit into parking spaces) of the huge cars of the late 1950s. Meanwhile, Vance Packard sought to expose the unethical business practices of automakers through their use of rapid style changes to fuel consumers' desire to own the latest model. An article in the New York Times spoke of a need for "rescue from the rubber-tired incubi," for moderation on the part of "the grand viziers of Detroit … because there is just not room on

1.24
A typical spread from one of Deborah Allen's articles about cars, demonstrating the "brilliant plumage" of the cars of 1955. "Guide for Carwatchers: Cars '55" by Deborah Allen, *Industrial Design*, February 1955, 82. Courtesy of F&W.

the streets or in the parking places," and suggested that "the system is exhausting the elements necessary for human life—land, air, and water."[162]

In panel discussions at the Museum of Modern Art and the International Design Conference at Aspen, and in articles in the design press, it was the aesthetics of car styling that was targeted—the use of "vulgar, superfluous, gadgety decorations, stripings or curlicues, accompanied by cheap color," as Raymond Loewy put it in 1951, or "borax," as MoMA's Director of Industrial Design, Edgar Kaufmann Jr. termed it in 1948.[163]

In art historian C. Edson Armi's view, MoMA, which excluded modern mass-produced American cars from its collection, "treated the car like an illegitimate child. After all, the primary function of a car's appearance was sales, and the 'philosophy' of its designers was likely to be a combination of power, fantasy, raw sexuality, and newness for its own sake—all basically abhorrent to the Bauhaus-oriented industrial arts establishment."[164] *Industrial Design*, by contrast, conducted comprehensive car design reviews in response to the automakers' annual changes, and can be seen as an emphatic example of the new type of criticism of popular, mass-produced design with which this chapter is concerned.

Despite her compelling coverage of the automobile industry for the magazine until the late 1950s, Deborah Allen is not well known as a design critic. She came into the profession through a series of chance encounters, rather than being driven by a mission. For four years at *Industrial Design* she wrote a series of sharp analyses of car design, and then stopped abruptly, due to the pressures of family life, never to be heard from again in a design context. Allen's oeuvre is well worth examination, however, since she reckoned with the design of cars, the most visible and profitable manifestation of American mass production, with a level of acuity and stylistic flair unparalleled among design critics of her time, and since.

Overall, Allen had little patience with the "expensive toys" she reviewed as a car critic.[165] She lived in New York, used public transport, and didn't even like cars that much. "It was hard to write about them because I thought they were senseless," she said of the exaggeratedly low-slung, long and streamlined cars of the period.[166] One review began: "In 1957, as far as we can make out, the American cars are as expensive, fuel-hungry, space-consuming, inconvenient, liable to damage, and subject to speedy obsolescence as they have ever been."[167] Allen's impatience with the stylistic flourishes of cars also comes through in her reviews. Of the 1958 Chevrolet she wrote: "The gull wing is as easy to identify and as annoying in its relationship to the rest of the car as all of GM's trademark tails." And to Allen, the "arbitrary whiplash" of another model's rear fender "is the final straw that makes one wonder what sense there is in any of these curves."[168]

Her mind changed, however, in early 1955, when riding into New York from Westport in a friend's Buick Century, as we saw at the beginning of this chapter. Inspired by her exhilarating experience of the car speeding along the coastal road, and her appreciation of the way her friend the driver inhabited its interior space, she wrote an uncharacteristically enthusiastic review. She referred to the Buick as a "slab on waves," demonstrating what she meant with accompanying diagrams. Allen was skeptical of the designers' desire to create an illusion of weightlessness, since the materials were actually very heavy. She observed that it was hard to believe in the "diaphanous" pretense of the Buick's heavy rear cantilever when you witnessed the effect upon it of a bump in the road. She wrote: "This attempt to achieve buoyancy with masses of metal is bound to have the same awkward effect as the solid wooden clouds of a Baroque baldacchino." But she went on to suggest that the beholder should "accept the romantic notion that materials have no more weight than the designer chooses to give them."[169]

The last line of the short review reads: "The driver sits in the dead calm at the center of all this motion; hers is a lush situation."[170] The depiction of a female driver refers to Allen's personal experience of this particular car, but also to the fact that most publicity shots supplied by car manufacturers featured women driving their cars. Manufacturers used women both to model the car and to acknowledge that women were key decision-makers in the purchase of family cars in the US; also, due to the postwar demographic migration to the suburbs, increasing numbers of women needed their own cars to perform household management tasks or get to work.[171]

The lyricism of the closing phrase, "hers is a lush situation," is achieved through the self-consciously poetic use of the third-person possessive pronoun, a set of circum-stances as the object, and the calculated evocation of the multiple meanings of the word "lush"—an adjective used to describe luxuriant vegetation, the state of being lavishly productive, and also something that is sensual and sumptuous. The phrase also conjures a novel image of a 1950s American woman, not trapped in the meaninglessness of her suburban existence as Betty Friedan and others portrayed her but, rather, calmly poised, in control of 5,000 pounds of metal, and embodying all the potential for productivity and growth evoked by the term "lush."

Industrial Design was run on a small budget. There was no money for Allen to go to Detroit for firsthand reporting, so she based her analyses on what she "saw on the road" and examination of the brochures the manufacturers sent her.[172] In this way she made use of art-historical techniques, such as comparison and type analysis, that she would have studied at Smith College and practiced briefly at the Metropolitan Museum of Art. Indeed, in a 1955 essay titled "Vehicles of Desire," Reyner Banham referred to Allen's "ability to write automobile critique of almost Berensonian sensibility."[173] Banham was referring to the American financier and art historian Bernard Berenson, who specialized in the Renaissance and whose analytical approach, codified in his essay "The Rudiments of Connoisseurship (A Fragment)," was highly influential in art history. According to the art critic Robert Hughes, Berenson pursued a "scientific" ideal of connoisseurship: "a system of discrimination based not on any special power of argument, still less on the icono-graphical or social meanings of art, but on meticulous observation of detail, sensitivity to style, and exhaustive comparison based on a retentive visual memory."[174]

Allen betrayed her art-historical training in another review, of 1955's brightly colored cars. She drew attention to the replacement of sheet metal, which had previously been used to convey speed, with that year's use of paint to describe "the more exaggerated effects of motion—a far more fitting medium for such impressionism."[175] And, in her appraisal of the 1955 Studebaker's "rakish" new body shape, she wrote in form-appreciative terms: "It is a stylish, Italianate combination of slow compound curves and sharply contrasting angles."[176] Again, in her review of the 1958 new Lincoln, we see more of her art historian's eye at work:

> American cars often look as if they were based on quick sketches rather than a careful study of form. At Ford, especially among the high-price cars, these sketches are apparently in clay: on Lincoln's side body, the sculptor's tool shows clearly in swift long lines, sharp edges, and concave modeling. This breeziness is slightly out of place in expensive hardgoods—with a little more time the sculptor would certainly have smoothed out the kick of metal ahead of the front wheel, the dust-catching ledge down the body, the extra metal at the back window. Furthermore, this sophisticated side-modeling conflicts with front and rear motifs that seem to be borrowed from below: sloping light mounts and chromed ovals recalling Edsel and Mercury and coy wings from the lowly Ford.[177]

Additionally, Allen's decision to crop photographs of cars to highlight certain features such as rears, bombs, posts, bulges, spears, saddles, speed lines, doors, and bumpers, and her meticulous assemblage of these images in pairs and typological groups, recalls the Wölfflinian technique of visual comparison so fundamental to the art history slide show.[178]

Such deployment of the visual rhetorical strategy of comparison and grouping was unusual in design criticism at that time. But by the late 1950s, companies advertising in *Industrial Design* were using similar techniques. In an advertisement for Rohm & Haas Plexiglas in the April 1957 issue, for example, eight cropped images of tail fins from various cars were shown in a grid over a spread with the tagline "What do they have in common?"[179] And in the same issue, an advertisement for Enjay Butyl rubber displayed all the rubber components of a car, just as Allen had done with zinc die castings in her review of 1957 cars a month earlier.[180]

In addition to her appreciation of the car as image, Allen's analysis also demonstrated a concern with the realities of its use. Her sensitivity to the ways in which people inhabited cars, and to how industrial design was experienced physically, differentiated her writing from ocular-centric, connoisseurial art criticism. She often drew attention to cars' safety hazards—the protruding rockets on the grilles, the sharp edges and knobs of the interior dashboards, and the poor visibility of wraparound windshields—and the cramped conditions of car interiors, especially the third man spots over the drive shaft "hillocks." Allen's discussion of use was not confined to ergonomics and functionality, however. She also took into account the mental and emotional qualities of driving. In a section of her

In absence of wheels, motion is suggested with decorations applied to body. Typical GM motif bevels Cadillac toward back.

Chrysler, usually conservative and functional, suddenly joins the opposition, slaps streams of chrome across '54 Dodge.

New Buick Super enlarges wheel openings to reveal sporty wire wheels. Chrome ignores structure, but is unusually modest.

Chrysler's experimental C200 suggests sportscar influence may help. Full quota of chrome is used to emphasize circle of wheels.

Sporty continental details are hard to adapt. Exposed spare blocks Nash trunk. Hard-to-wash wire wheels, used to cool brake drums on racing car, are simulated on heavy Cadillac convertible.

Lowslung car crowds passengers, is hard to enter. GM suggests swivel chairs on experimental Pontiac 5-man sedan. Buick station wagon has no third seat.

'48
'51
'55
'54

Sleek ram, Dodge radiator motif, illustrates design progress since 1946.

Powerful bumpers also show progress. Dodge Firearrow and Packard Panther share new crusty look. GM tries gentlemen's approach on experimental Wildcat, sticks to "bombs" on real Cadillac.

To what extent are the new cars really better than the old ones?

One of the American manufacturer's main criticisms of the European sports car has been that it is cramped and inadequate for family use. To the extent that family cars are made to emulate the racier lines of sports cars, they too become less adequate for utilitarian needs.

Since American automotive engineering has always emphasized comfort over safety, the construction of American cars has not been developed to a point commensurate with their high power and performance. A softly sprung car is not safe at high speed, nor is the conventional steering gear, designed to lighten the effort of turning the wheel, quick enough in an emergency. The American type of car, with its excessive weight, size, speed, and power, is potentially the most destructive in the world.

The streamlined shell that encloses the American car, while it gives an air of great size and power, renders the working parts inaccessible. The engine is within easy reach only when the hood is sloped. Covers over the wheels, aside from hiding the car's most emphatic shape, are a mechanical nuisance.

Such complaints are old, however, and if they mattered Detroit would answer them. The real point, in the '54 cars, is styling, and the real criticism is that, even in Detroit's terms, it's not wonderful. One would think a car might be big and shiny *and* beautiful all at once.

1.25
A spread from one of Deborah Allen's articles about cars, demonstrating her typological groupings of car features such as fronts and rears, "Guide for Carwatchers: Cars '55" by Deborah Allen, *Industrial Design*, February 1955, 84–85. Courtesy of F&W.

1955 review devoted to the positioning of the Plymouth's posts (the vertical structural elements that support the roof of a car), she concluded: "At GM a post isn't a post, it is a design on your emotions, and if it defies purist logic, it nonetheless succeeds in its real aim, which is purely psychological." And of the 1955 Buick, she wrote: "But when the driver gets into the car ... something else begins to operate. In the Buick she is couched at just the right point among the flattering curves, and her distance from the windshield gives her an air of command that may do more for her driving than a clear view of the road."[181] In a special feature titled "Cars '56: The Driver's View," Allen led with a picture of a steering wheel and dashboard in which three disembodied white-gloved hands manipulated the car's "appalling number of gauges, controls, and push-button devices," which could include record players, air-conditioning, ashtrays, antenna, and convertible top controls. The article made typological comparisons between features like speedometers and crash features, using cropped photographs gathered in tight juxtapositions and a listed taxonomy of all the "Watch" and "Work" functions of the car. In her introduction she opined: "Yet logic and legibility are only one part of dashboard design. A second challenge—and often it seems the major one—is psychological. As a nerve center of the car, the dashboard explains and advertises its performance and builds up the pleasure and excitement of driving. Like most psychological problems, this one is complex: the car must generally look powerful and heavy yet fast and maneuverable, loaded with conveniences yet simple to master, safe yet daring, lush yet sporty."[182]

Furthermore, Allen's writing shows that she also understood the interrelated economic processes of manufacture, retail, and distribution. She tracked sales figures and made predictions about a model's commercial success. She explained technical aspects of car production with clarity and precision, using diagrams to supplement her written description. In her review of the 1958 Chevrolet, for example, she wrote: "To achieve the lowness of its competitors, Chevvy uses a new frame that seems to provide good interior space. ... Rather than a box frame or an x-frame, this is an 'hour-glass' frame that concentrates structure at the driveshaft, where there is a hump anyhow. In place of the heavy side rails that brace the usual x-frame, Chevvy has light rails attached to the body rather than to the frame."[183]

Informed by art-historical study and literary flair, Allen's writing was also tempered by lived experience and technical knowledge, and applied to human interaction with cars as well as the mechanics of their economic exchange. Allen was, as she put it, deeply interested in car design, not on moral grounds—"we can't say this is wrong, any more than Eve was wrong"—but simply because cars at the time were "the most unavoidable, costly, and popular example of industrial design on the American market, and of all popular American products they are the most aesthetic in concept and purpose."[184]

Allen struggled to balance the pressures of running a large family and maintaining an editorial career. She and her husband had moved to Washington, D.C., and she commuted to New York for some time, taking a magazine's worth of copy to edit on the train,

but finally bowed out in 1957, leaving the editorship in Thompson's hands (although she would continue as a consultant to the magazine for a few more years).

As we have seen, Allen's work from this period had an unexpected second life in Richard Hamilton's art work, specifically his "Hers is a Lush Situation" series in which the lipstick-red mouth of a bodiless driver hovers above a diagrammatic inventory of Detroit styling features including visor-hooded headlamps, chrome spears, tail fins, speed markings, and a CinemaScope windshield, details which Hamilton had gleaned from Allen's car reviews. Despite her own disillusionment with her subject matter and her rejection of the medium she was so skilled in, therefore, Allen's writing transcended, or at least escaped, its genre and made a curious voyage across continents, disciplines, and contexts to live on in the canons of British, and international, art.

Reyner Banham, too, found in Allen's writing inspiration for his own appraisals of cars, and more generally for his desire to develop a new mode of writing about the expendable, mass-produced materiality of popular culture. In 1955, he declared excitedly of Allen's Buick review: "This is the stuff of which the aesthetics of expendability will eventually be made."[185] He applauded Allen's writing for its ability to channel the vitality of the Detroit body stylists themselves, to approximate "the sense and dynamism of that extraordinary continuum of emotional-engineering-by-public-consent which enables the automobile industry to create vehicles of palpably fulfilled desire."[186] Banham saw the body stylists of the automobile industry, vilified by most other design writers both in the US and the UK, as providing essential arbitration between industry and the consumer. Their work would become a key reference point for him in developing his new literary arsenal for dealing with popular culture, in his "attempt to face up to Pop, as the basic cultural stream of mechanized urban culture."[187]

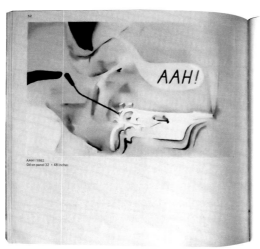

Although Banham did not learn to drive until 1966, preferring the Moulton bicycle as a mode of transport through London's streets, and regarding "auto-addicts" as "an ugly mob," he found in cars subject matter that suited his knowledge of engineering and appreciation of popular culture.[188] In the 1960s, during travels to the United States, and possibly inspired by Allen's writing, he began to appreciate the physical experience of driving, writing of negotiating Los Angeles freeways in ecstatic terms: "To drive over those ramps in a high sweeping 60-mile-an-hour trajectory and plunge down to ground level again is a spatial experience of a sort one does not normally associate with monuments of engineering—the nearest thing to flight on four wheels I know."[189]

PART THREE: DEVELOPING AN AESTHETICS OF EXPENDABILITY: REYNER BANHAM'S CRITICAL DESIGN WRITING, 1955–1961

"Vidiot" consumers

In the late 1950s and early 1960s, Reyner Banham was preoccupied with formulating a new type of critical writing equipped to tackle popular, mass-produced, expendable product design. Banham believed that the modes and values of design criticism as it had been conducted were distilled from the precepts of Modernist architecture, and thus were out of date and insufficient for any convincing appraisal of the expendable products of a throwaway economy. New criticism would require new diction, metaphors, syntax, methods, purpose, values, and readerships. It would also require a sensitivity to the products under consideration, and an empathy with the concerns of their consumers.

In the 1963 "Who Is This Pop?" article we glimpsed at the beginning of this chapter, Banham had introduced the term "Vidiot," which he characterized as someone "trained to extract every subtlety, marginal meaning, overtone or technical nicety from any of the mass media," and thus in this term he conflated his own role as critic with the knowing consumer he represented.[190] Later in the piece he claimed that there is no such thing as an "unsophisticated" consumer: "all consumers are experts, have back-stage knowledge of something or other, be it the record charts or the correct valve timing for doing the ton."[191] In an article which surveyed the landmarks and influences that he felt had shaped industrial design criticism of the past decade, he highlighted Consumer Research, and the way in which "the formal recognition of a specific consumer viewpoint in relation to industrial design" had emerged as "one of the more important new factors."[192]

Banham advanced his argument for a new type of design writing in several articles of the period by tracing the historical lineage of industrial design criticism, critiquing contemporaneous writing and the influence of design institutions, and by experimenting himself with the nascent form. Assembled together, these fragments of various articles constitute Banham's statement of practice as a product design critic.

　　　　　　　　　　　　　　　"A THROW-AWAY ESTHETIC"

"Goodies," not "good design": Banham's subject matter[193]

Banham identified his subject matter as the kinds of new, cheap, mass-produced, often "flashy and vulgar" products that were increasingly figuring in working-class people's lives, as "physical and symbolic" consumables. He referred to these objects as "goodies." A goodie's "sole object was to be consumed," he observed, in the article "Who Is This Pop?," referring readers to the passage in Vladimir Nabokov's novel *Lolita*, in which the narrator buys the prepubescent Lolita "four books of comics, a box of candy, a box of sanitary pads, two cokes, a manicure set, a travel clock with a luminous dial, a ring with a real topaz, a tennis racket, roller skates with white high shoes, field glasses, a portable radio set, chewing gum, a transparent raincoat, sunglasses."[194] Banham was more specific than Nabokov in his characterization of goodies. They had to be brand names, carefully sourced, and then edited and curated into "a fairly compact package" or kit. Things like "genuine Brand-X cigarettes, Japanese wrestling magazines, foreign paper-backs from Krogh and Brentano's" were gathered and presented like urban archaeological finds. The artist John McHale, for example, had spent a year in 1955 as a visiting student at Yale University and returned to London with a trunk full of American magazines such as *Life*, *Look*, and *Esquire*, and other ephemera, that provided the sources for Banham and Hamilton's studies of American white goods and cars. Later, in 1963, the architect Warren Chalk assembled the components of an imagined contemporary urban *flâneur's* "Living City Survival Kit," in his image for the catalog for Archigram's "Living City" exhibition at the ICA. The image, which includes sunglasses, instant coffee, a packaged Wonderloaf, a box of tissues, lipstick, an issue of *Playboy* magazine, and an Ornette Coleman record, referenced a newly emerged urban lifestyle made up of disposable, instant, pocket-sized, mass-produced consumer goods, with no discernible ancestry as object types. The dilemma for Banham, and other like-minded writers, was how to approximate this condition and these appurtenances through language. He turned to his provincial upbringing for answers.

Banham, born in 1922, was raised in Norwich in the eastern British county of Norfolk, the son of a gas engineer. He trained in aeromechanical engineering at Bristol Technical College, focusing on management training, then worked at Bristol Aeroplane Company as an engine fitter. After the war he returned to Norwich, where he wrote reviews of art exhibitions for local newspapers such as the *Eastern Evening News* (as P.R.B.) and the *Eastern Daily Press* (as Reyner Banham), and enrolled in an adult education art history course taught by Helen Lowenthal. With Lowenthal's assistance, and after learning German (the language required for entry), he was admitted to the Courtauld Institute in London to study architectural history. When he earned his BA in 1952 he began to study with the German-born architecture critic and historian Nikolaus Pevsner, working on a PhD about "developments in architectural form and ... architectural thought in an industrialized epoch," which he published in 1960 as *Theory and Design in the First Machine Age*.[195]

While studying, he also worked as a part-time literary editor at the *Architectural Review* and attended the Independent Group meetings at the ICA as an organizing member and recorder.

Through the IG discussions Banham realized that his working-class, rural upbringing—a disadvantage at the Courtauld Institute, and in the art history profession more generally—was, in the pluralist atmosphere of Pop, actually an asset to be leveraged. The usual trajectory for a Courtauld graduate, according to his widow Mary Banham, was to go and work in a provincial gallery or museum, with a view to returning to London after a few years. She believes that Anthony Blunt, the director of the Courtauld, and Pevsner helped him to circumvent this route, because Banham was already an accomplished journalist and did not want to leave London, but mainly, she suspects, because "he was not a gentleman and said what he thought."[196] Banham became increasingly comfortable with the fact of his working-class background, using it to his advantage, and in 1964 claimed that it gave him "a right to talk about certain subjects."[197]

Banham was keen to locate himself at the wellspring of Pop ideas—as someone who had "helped to create the mental climate in which the Pop art painters have been able to flourish." In articles he often emphasized his working-class roots and those of most of the Independent Group members who, he said, were all brought up "in the Pop belt somewhere," all-knowing consumers of American films and magazines in an inevitable rather than a studied way.

The following autobiographical passage is typical of several similar instances in which Banham cements a narrative of his coming of age through the practice of knowing, pluralist consumption, in which a new "scale of values" embraced both comics and literature: "I have a crystal clear memory of myself, aged sixteen, reading a copy of *Fantastic Stories* while waiting to go on in the school play, which was Fielding's *Tom Thumb the Great*, and deriving equal relish from the recherché literature I should shortly be performing and the equally far out pulp in my hand. We returned to Pop in the early fifties like Behans going to Dublin or Thomases to Llaregub, back to our native literature, our native arts."[198] By training his gaze on mass-produced goods and the manifestations of popular culture, Banham made a political statement that clashed with most design commentary to date, which usually excluded this material, using criteria of restrained aesthetics and durability as its selective filter.

Working as a freelance writer for various publications, rather than having to toe any institutional party lines, Banham was freer than most design commentators of the period to explore different topics, stances, and writing styles. In 1956 he wrote about industrial design and "the common user" for the *Listener*, and with his "Not Quite Architecture" column for the *Architects' Journal*, begun in 1957, and his *New Statesman* column on architecture, technology and design, begun in 1958, he experimented with broadening his field to include reviews of science fiction and blockbuster films, and industrial design or the themes that framed it, such as the retreat of the Italian influence in British society. By the

"A THROW-AWAY ESTHETIC"

mid-1960s, with a weekly "Arts in Society" column in *New Society*, Banham was known for his coverage of mass-market goods, particularly their packaging, which he saw as enabling "the latest and most sophisticated types of design into domestic environment," via frozen foods, LP records, and paperback books.[199]

"Many, because orchids": Banham's critical values[200]

The new subject matter that Banham had identified demanded a corresponding shift in values that grated with the establishment view of design. A critic of serially produced popular product design would have to grasp the implications of expendability, decoration, and manufacturing and marketing processes. He would also have to have the ability to intuit the desires of the kind of knowing consumer who could distinguish "stainless from spray chrome at fifty paces," as well as the worldview of the designer.[201]

Banham dismissed what he saw as a century of thinking about designed products sustained by "a mystique of form and function under the dominance of architecture," and misled by a confused idolization of simplicity and standardization. Inspired by automobile designer Jean Gregoire's observation that the European Bugatti engine, with its carefully hidden wiring and accessories, was in fact less beautiful than American engines where the manifolds were clearly seen and easy to access for repair purposes, Banham compared a Bugatti engine with a Buick V-8: "The Bugatti, as Gregoire noted, conceals many components and presents an almost two-dimensional picture to the eye, while the Buick flaunts as many accessories as possible in a rich three-dimensional composition, countering Bugatti's fine art reticence with a wild rhetoric of power."[202] Summarizing the appeal of the Buick, he enumerated the following qualities: glitter, bulk, three-dimensionality, deliberate exposure of technical means, ability to signify power, and immediate impact. To Banham, these qualities represented the antithesis of fine-art values and fulfilled instead the literary critic Leslie Fiedler's definition of Pop Art articulated in an essay on comic books in *Encounter*, which Banham appreciated.[203] Banham quoted Fiedler, who had written that although contemporary popular culture differs from folk art, in "its refusal to be shabby or second rate in appearance, its refusal to know its place," it is not designed to "be treasured, but to be thrown away."[204]

The design historian Gillian Naylor, among others, has pointed out the importance of the Futurist movement for Banham's intellectual development—his belief in technological invention, the need to study mass-produced design, an anti-permanence philosophy, and an embrace of what Italian Futurist Umberto Boccioni had called the "anti-artistic" manifestations of our epoch, can all be traced back to his research into the Futurists for his revisionist history of modern architecture, *Theory and Design in the First Machine Age*.[205]

In his 1955 article "Vehicles of Desire," Banham bemoaned the fact that Platonic ideals of permanence, more befitting to architecture, were still being used to measure value in industrial design: "We are still making do with Plato because in aesthetics, as in most other things, we still have no formulated intellectual attitudes for living in a throw-away economy."[206] He continued: "We eagerly consume noisy ephemeridae, here with a bang today, gone with a whimper tomorrow—movies, beach-wear, pulp magazines, this morning's headlines and tomorrow's TV programmes—yet we insist on aesthetic and moral standards hitched to permanency, durability and perennity."[207]

Banham continued his theme in "Design by Choice," published in July 1961 in *Architectural Review*. He opined that while the Modern Movement held sway in the early twentieth century, architects such as Voysey, Lethaby, Muthesius, Gropius, Wright, and Le Corbusier, and writers influenced by them such as Edgar Kaufmann Jr. in the US and Herbert Read in Britain, directed the production and discussion of industrial design. By 1961, however, Banham noted, architects had ceded control of the discussion to "theorists and critics from practically any other field under the sun":

> The new men in the USA, for instance, are typically liberal sociologists like David Reisman or Eric Larrabee; in Germany, the new men at Ulm are mathematicians, like Horst Rittel, or experimental psychologists like Mervyn Perrine; in Britain they tend to come from an industrial design background, like Peter Sharp, John Chris Jones, or Bruce Archer, or from the Pop Art polemics at the ICA like Richard Hamilton. In most Western countries, the appearance of consumer-defence organizations has added yet another voice, though no very positive philosophy.[208]

In these newly configured circumstances, opinion on industrial design was fractured and eclectic, and served the ideological purposes of each commentator. Banham attempted to convince the readership of the *Architectural Review* that they needed to understand product design, and that they would need a guide in such unfamiliar territory. By inserting numerous hints of his knowledge and ability to translate jargon terms such as "Detroitniks" and "hidden persuaders," Banham prepared the way for his own indispensable role as guide.

Negotiating "the thick ripe stream of loaded symbols": Banham's methods[209]

In Banham's view, even though the material qualities of products had changed, the basic "problem" of industrial design had remained the same—"it is still a problem of affluent democracy, where the purchasing power of the masses is in conflict with the preferences of the élite."[210] Banham outlined a new and commercially focused role for the product critic, as partner of the designer, which is "not to disdain what sells" but to help industry determine "what *will* sell." Part of this role involved selling not just the product to the consumer, but also the consumer to capital: "Both designer and critic must be in close touch with the dynamics of mass-communication. The critic, especially, must have the ability to sell the public to the manufacturer, the courage to speak out in the face of academic hostility, the knowledge to decide where, when and to what extent the standards of the popular arts are preferable to those of the fine arts. He must project the future dreams and desires of people as one who speaks from within their ranks."[211]

By urging critics to get closer to the design industry and to participate more actively in its manipulation of popular desire, Banham took a contrary stance, and one that identified "academic hostility" as the primary impediment to progress, rather than manufacturers or designers.

Some of the academic hostility came from the direction of the Hochschule für Gestaltung Ulm in Germany, under the directorship of Tomás Maldonado, which Banham had characterized as the "cool training ground for the technocratic elite."[212] Banham was invited to visit the school in March 1959, and he delivered two lectures, "The Influence of Expendability on Product Design" and "Democratic Taste," which were afterward "heatedly discussed."[213] In fact, it is hard to picture a setting more antithetical for Banham to present his ideas on the virtues of ephemerality and the idiosyncrasies of public taste. Ulm's pedagogical philosophy under Maldonado was highly scientific and technological, and underpinned by functionalism. Espousing Frankfurt School arguments, Maldonado drew attention to what he saw as Banham's mistaken assumption that Detroit car styling was an expression of the people, when in fact it was a calculating marketing exercise designed cynically by large corporations. He wrote: "I am not much convinced that the aerodynamic fantasies of Vice Presidents of Styling have much in common with the artistic needs of the man in the street."[214]

Maldonado, an anticapitalist design theorist committed to a rational approach to design, saw Banham's argument as fundamentally flawed. Banham's point, however, was that to truly understand industrial design as a critic, one needed to get close to the sources of both manufacture and consumption—to report from the ground, rather than to philosophize from a distance. What does appear contradictory in Banham's argument is his requirement that a critic of popular product design should be an ally of the designer and help serve up the consumers on platters to the industrial complex, while also representing the emotional desires of the knowing consumer.

Banham set out a method for critical analysis in the new conditions of expendability, which would take into account a product's content, its symbolism, and the popular culture it spoke to. The proper criticism of popular product design depended, he opined, on "an analysis of content," "an appreciation of superficial rather than abstract qualities," and an ability to see the product as "an interaction between the sources of the symbols and the consumer's understanding of them."[215] He explained how a critic "must deal with the language of signs." Improved criticism was contingent upon "the ability of design critics to master the workings of the popular art vocabulary which constitutes the aesthetics of expendability."[216]

Banham highlighted a sample of Deborah Allen's writing about cars in *Industrial Design*, discussed earlier in this chapter. He regarded Allen as one of the few commentators equipped to write about cars and "the thick ripe stream of loaded symbols" with which stylists adorned them.[217] Seeking an alternative to architecture with which to compare cars, Banham lit upon comics, movies, and musicals as the nearest point of reference, for these products bore "the same creative thumb-prints—finish, fantasy, punch, profession-alism, swagger." Top body stylists, he argued, were looking in the same direction. They used symbolic iconographies "drawn from Science fiction movies, earth-moving equipment, supersonic aircraft, racing cars, heraldry, and certain deep-seated mental dispositions about the great outdoors and the kinship between technology and sex."[218] Deploying such popular visual references, the body stylists were able to mediate between industry and the consumer, and to say something of "unverbalisable consequence to the live culture of the Technological Century."[219] It was this ability of the Detroit body stylists to conduct a "repertoire" of styling details, to "give tone and social connotation to the body envelope," and to connect to a "live culture" that Banham sought to capture and make "verbalisable" through his own writing.

"Boeing along to Honolulu": Banham's language[220]

Banham's most significant and enduring contribution to a new form of product design criticism is to be found in the diction he introduced to design discourse. The broader cultural project of using language to approximate the contours of a Pop sensibility was already underway in the literary forays of authors such as Anthony Burgess, especially in his novels *Nothing Like the Sun* and *A Clockwork Orange*. Literary critic John J. Stinson observed: "The art that Burgess gives us is, in fact, very much akin to that of the Pop Artists of the graphic arts, chiefly in the fact that the countless mundane objects he gives us come very near themselves to being the subject matter, although also as in the graphic arts, they are superinflated (in Burgess by a bursting sort of neo-Jacobean language) so as to bring us to new perceptual and ontological levels of awareness."[221]

Burgess himself later observed: "By extension of vocabulary, by careful distortion of syntax, by exploitation of various prosodic devices traditionally monopolized by poetry, surely certain indefinite or complex areas of the mind can more competently be rendered."[222] In nonfiction writing, too, American journalists such as Tom Wolfe and Gay Talese were exploring a new immersive approach to storytelling, saturated with technical detail, allusion, extensive passages of dialogue, and imagined scenarios, which Wolfe later dubbed the "New Journalism."[223]

Banham transplanted neologisms, the rhythm and diction of contemporary vernacular dialogue, the language and brand names of commercial culture, and poetic phrasing to the context of design writing. Consider just one of his sentences: "The New Brutalists, pace-makers and phrase-makers of the Anti-Academic line-up, having delivered a smart KO to the Land-Rover some months back, have now followed it with a pop-eyed OK for the Cadillac convertible."[224] Here Banham hyphenated words to make new ones (pace-setters, phrase-makers, pop-eyed), emphasizing the condensed information-packed impression of the sentence. He used the colloquial abbreviations KO and OK in a pleasingly symmetrical and palindromic shorthand for evoking his perception of a change in taste (the British establishment, as represented by the sensible Land-Rover, was given a "Knock Out," while the excesses of Detroit car styling symbolized by the Cadillac were given approval). Through such playful linguistic devices Banham began to work out a distinctive writerly voice capable of engaging with the vitality of popular culture on its own terms.

The women on the bus: Banham's readers

Banham, who between 1958 and the late 1970s was writing hundreds of weekly columns, knew very well the pressures of writing to deadlines and directly into the fast-flowing current of contemporary culture. His articles about design can be considered as expendable as the topics he was writing about. Reflecting on the journalistic aspect of his oeuvre, he wrote: "the splendour (and misery) of writing for dailies, weeklies, or even monthlies, is that one can address current problems currently, and leave posterity to wait for the hardbacks and PhD dissertations to appear later."[225] Banham's belief in expendability extended to the record of his own work. He burnt all of his papers in 1976 before he moved his family to Buffalo, New York. "He wasn't interested in posterity," his widow Mary observed.[226]

Banham was a dexterous and witty writer who wrote out in longhand on foolscap paper preparatory versions of his articles before typing them up. He would show them to Mary, an art teacher by training, who, in addition to doing architectural drawings for his articles for the *Architectural Review*, said that she performed for him the role of "the woman on the bus, or everyday reader."[227] Mary said she helped him "break down his long sentences" and made him explain technicalities, "because he wanted to introduce what he was interested in to as big a public as possible."[228]

Through publishing in popular mainstream publications, Banham made the tools of criticism available to his readers so that more people could apprehend the designed environment that surrounded them. He used the iconographic methods of art history he learned as a student at the Courtauld Institute, in which one focused on the identification, description, and interpretation of the content of images, but he applied them to designed objects and phenomena that lay beyond art or even architecture criticism's regular territory—he took criticism out into the field in order to "participate in the extraordinary adventure of mass-production."[229]

CONCLUSION

The March 1960 issue of *Industrial Design*, guest-edited by Jane Thompson as her last effort for the magazine as a consulting editor, was an anthology of 40 articles and excerpts, written by foreign critics gathered from design magazines in Italy, France, Germany, Norway, Sweden, India, and England. Thompson wanted to explore the differences between European and American design, which she saw as being at different stages of development in terms of their large-scale production and competitive marketing. Freed from the responsibility of being the editorial figurehead of a magazine founded to promote

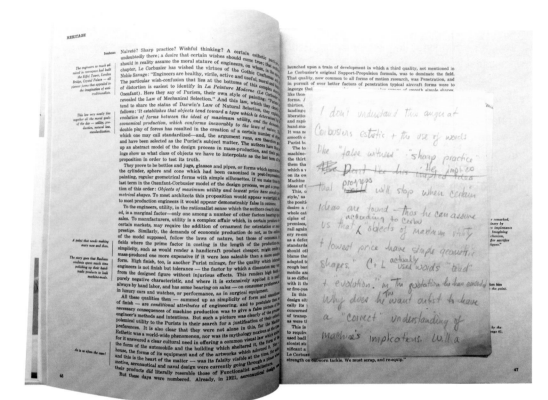

1.28
Spread from the reprint of Reyner Banham's "Machine Esthetic" article in *Industrial Design*'s special international issue in 1960, showing Jane Thompson's commentary in the margins and Deborah Allen's handwritten notes in response to Banham's argument. Courtesy of F&W and Oliver Allen.

"A THROW-AWAY ESTHETIC"

the interests of American designers to industry, in this issue Thompson was able to introduce more critical content than she had thus far.

Through her selection of such a variety of voices, the text-heavy nature of the issue, the complex layout of the magazine which incorporated her chatty marginalia and responses from writers to particular claims in articles set alongside the appropriate passages, Thompson created the feeling of a live debate in action, and a snapshot of international design discourse in the late 1950s as filtered through her editorial viewpoint. In her introduction she observed:

> Overseas [the designer] puts out fewer products and more words than his busy American counterpart. … But is this really for the lack of time and thought? Doesn't this really go back to the traditional belief, as old as the depression-born profession itself, that to sell itself to business, industrial design had to adopt the standards of business, and cut itself off from the American arts? Our self-willed isolation has had curious effects, among them the lack of a critical tradition among designers and the lack of any active school of professional critics who support the designer in his search for valid expression and purpose. There are many ramifications to this critical void, but they boil down to this: US industrial design itself has not believed in criticism or accepted it, because it grew up on business' belief that you can't criticize design if it sells, daren't criticize it for fear of harming sales.[230]

Among the featured articles were Banham's "Industrial Design and Popular Art," republished with the new title "A Throw-away Esthetic," and an excerpt from Hamilton's "Persuading Image." In this new context, these articles feel incongruous in their lack of concern for the social issues that were beginning to absorb intellectual design culture. Banham's piece had been written five years previously and Hamilton's article, although it had been published in *Design* magazine only the month before, looked back to the mid-1950s in its references. Industrial designer Don Wallance pointed out the anachronistic nature of the articles in a letter published in a subsequent issue. Referring to Banham's piece, Wallance wrote: "Some of our friends having belatedly embraced the techniques of mass marketing are not content merely to enjoy its economic benefits, but are impelled to idealize and institutionalize its esthetic consequences." He went on to point out that this "is at a time when many thoughtful Americans such as John Galbraith, Walter Lippmann and C. Wright Mills are questioning the economic and social premises of the Big Sell that underlie Mr. Banham's throwaway esthetic."[231]

Wallance's observation suggests a disconnection between Hamilton and Banham's fascination with American consumer culture of the late 1950s and the emergent concerns of some American designers. By 1960 a new more serious, anxious, and morally driven species of design criticism was beginning to take shape that called for accountability in the design profession and its associated industries.

Of the three "thoughtful Americans" mentioned by Wallance, C. Wright Mills, a Marxist professor of sociology at Columbia University, was the most directly critical of design, and designers' complicity in eroding left-wing values through their role in the misleading conflation of culture and commerce. His ideas on design's complicity with the "cultural apparatus" had been aired to the design community at the International Design Conference at Aspen, and in the pages of *Industrial Design* magazine in 1958. Wright Mills questioned the way in which people's experiences were increasingly prefiltered by designers: "The world men are going to believe they understand is now in this cultural apparatus, being defined and built, made into a slogan, a story, a diagram, a release, a dream, a fact, a blue-print, a tune, a sketch, a formula; and presented to them." He posited that by squandering their responsibility as "observation posts" and "interpretation centers," designers were succumbing to commercial imperatives "which use 'culture' for their own non-cultural—indeed anti-cultural—ends."

Wright Mills identified planned obsolescence as the economic environment in which "the designer gets his main chance": "The silly needs of salesmanship are thus met by the silly designing and redesigning of things. The waste of human labor and material becomes irrationally central to the performance of the capitalist mechanism. Society itself becomes a great sales room, a network of public rackets, and a continuous fashion show."[232]

This article provoked several responses among *Industrial Design*'s readership. Fred Eichenberger, Assistant Professor of Design in the College of Applied Arts at the University of Cincinnati, wrote to commend the piece and to underline its moral message:

> It seems to me that the heart of Mills' proposed book is the consideration of public and private morality. We are all familiar with the statements of aims and ethics published by the various professional societies of design. These have to do mainly with the designer's working relationships, his obligations to his client, and his attitudes towards other professionals. Now this too is morality, but of a very specific sort. The kind of morality I mean is concerned with the way our efforts affect the larger society. In a world of exploding populations and exploding nuclear devices, of contracting natural resources, in a world in which urbanization and supra-nationalism are making enormous advances, all of us must, as never before, question the consequences of our actions.[233]

A growing body of thinkers was exposing the social, psychological, and physical dangers of planned obsolescence, public relations, motivation research, car design, waste, litter, and the lack of attention to Third-World poverty.[234] Such arguments were directed squarely at the very position that Banham and Hamilton had celebrated in their writing, and that *Industrial Design* magazine had nurtured so carefully in its editorial premise, which was to be "on the designer's side."[235] The idea that a design critic should be a designer's ally was beginning to crumble.

It was as if two tectonic plates of design criticism—one driven by a need to shake up old establishment values and to extend "the long front of culture" on its own new stylistic terms, the other directed by social, political, and moral concerns and in some cases recommending a return to the old values—were grating past one another as they headed in different directions. The two variants of criticism shared the same subject matter—cars and white appliances—but their motivations, arguments, style of language, and points of origin were profoundly different.

Banham stood his ground. In his "Design by Choice" article of 1961 he gave "the new men in the USA ... typically liberal sociologists" short shrift: "Lash-up formulations of this sort are, of course, only ad hoc intellectual structures and should be neatly put away when they have done the job for which they were assembled. Thus, a narrowly Stalinist frame of reference, rigidly maintained beyond its last point of utility, has resulted in the sterility and subsequent disappearance of radical left-wing design criticism in Western democracies, and leaves intelligent sociologists, like Richard Hoggart, apparently sharing the opinions of an 'Establishment' that they otherwise despise."[236] His debunking was focused in particular on Vance Packard. Under the heading "Alarmist Literature," he wrote:

> In the 1950s the shortcomings of some aspects of product design became a subject for sensational journalism which—in some cases—contained an element of serious warning. The most prolific of these professional Jeremiahs was the American writer Vance Packard, whose book *The Hidden Persuaders* drew attention to the social consequences of motivation research. His subsequent works *The Status Seekers* and *The Waste Makers* continued variations of the same theme of social enquiry into design, but began to suggest that he had fallen victim to the very situation against which he was protesting: his elevation to the best-seller list involved him in the dynamics of the mass market and more or less committed him to bring out a "new model" every other year.[237]

Banham would have plenty more to say about Pop and popular culture in the 1960s, and he continued to deploy his newly formed aesthetics of expendability on the explication of product design. Meanwhile, the wider climate of opinion was shifting away from a celebration of popular culture and technological progress toward a more questioning approach with regard to the social and environmental consequences of a disposable product design culture. Such concerns would force themselves onto the main stage of design discourse when, as we shall see in chapter 2, students and environmental activists disrupted the proceedings of the 1970 International Design Conference at Aspen, and Banham, acting as moderator, was confronted with a vehement backlash against the values of expendability, excess, and surface styling that he and others had spent the late 1950s and early 1960s endorsing so personally, persuasively, and poetically.

2.1
Still from *IDCA '70* showing members of the IDCA board
and their wives gathering for cocktails on the first
evening of IDCA 1970. Dir. Eli Noyes and Claudia Weill,
IDCA, 1970. Courtesy of Eli Noyes.

"Conflicting Definitions of Key Terms":
An Ecological Protest at the
International Design Conference in Aspen,
1970–1971

It was a perfect June evening in Aspen, a former silver mining town 8,000 feet up in the Rockies. As the sun began to dip behind the snow-capped mountains on this first day of the 1970 International Design Conference at Aspen (IDCA), a group gathered for a cocktail party. The men were dressed in plaid jackets and ties and their hair, if they still had any, was close-cropped and gray. Their wives, who had been offered a reduced conference fee, so that the weeklong conference might be combined with a family vacation, wore cocktail dresses, large sunglasses and pearls, and their hair, curled and set, barely moved in the breeze that ruffled the nearby aspen trees.

Drinks were set out on the terrace of one of the modernist houses in the Aspen Meadows complex designed by Herbert Bayer. The Austrian émigré and consultant to Container Corporation of America, who had moved to Aspen in 1946, was there, dapper in his suit and cravat, suntanned and still handsome at 70. Also sipping gimlets were other board members in charge of the IDCA: Saul Bass, the Los Angeles-based graphic designer; Eliot Noyes, design director at IBM and president of IDCA since 1965; and George Nelson, design director at the high-end office furniture firm Herman Miller. These men had been trained as artists and architects, but had helped to define the American graphic and industrial design professions in the 1940s and 1950s. Their careers had flourished in the postwar period of economic expansion, and were tied to the rise of a consumer society. Now in middle age, they held prominent positions within both the professional design community and the flagship corporations of the day. As they laughed, clinked glasses, and patted one another on the back in collegial amiability, it certainly looked as if these representatives of the American design elite were enjoying the fruits of their labors.

Meanwhile, in the meadow beyond this midsummer cocktail party for the cognoscenti of modernist American design, a very different scenario was taking shape. Milling about outside a big white concert tent, designed by Bayer, where the conference would be held, were student designers and architects, some of their young teachers, and a number of art and environmental collectives, many from Berkeley, California, who had made the 1,000-odd-mile journey to Colorado in chartered buses. With their waist-length hair, beards, open-necked shirts, bandanas and jean jackets, these new arrivals signaled both their adherence to an alternative lifestyle and set of values (for which

Berkeley was the unofficial American capital) and their distance from the largely East Coast conference organizers.

Among them were members of the San Francisco media collective Ant Farm, who, by 1970, were well known for their advocacy of a nomadic lifestyle, their use of inflatable structures as the setting for free-form architectural performances, and their experiments with video as a vehicle for critique. In a biographical statement Ant Farm had characterized themselves as "an extended family ... of environmentalists, artists, designers, builders, actors, cooks, lifers and an inflatable named Frank; war babies, television children, *Rod & Custom* subscribers, university trained media freaks and hippies interested in balancing the environment by total transformation of existing social and economic systems."[1]

Since the theme of the conference this year was "Environment by Design," several representatives of environmental action groups were also gathering, invited by Sim Van der Ryn, an assistant professor of architecture at the University of California, Berkeley, and founder of the Farallones Institute, a research center for studying environmentally sound building and design, and low-technology solutions to waste management.

Among those also invited by Van der Ryn were: Michael Doyle, founder of the Environmental Workshop in San Francisco; the People's Architecture Group; Steve Baer, the Albuquerque solar energy enthusiast who founded Zomeworks and developed many of the housing structures for communes such as Drop City and Manara Nueva; and Cliff Humphrey, founder of Ecology Action, originator of the first drop-off recycling center in the US, whose Berkeley commune, known for smashing and burying cars, had just been featured in a *New York Times* magazine cover story. The cover image showed Humphrey pushing a bandaged globe in a baby stroller.[2]

2.2
Still from *IDCA '70* showing members of Ant Farm at IDCA 1970. Dir. Eli Noyes and Claudia Weill, IDCA, 1970. Courtesy of Eli Noyes.

"CONFLICTING DEFINITIONS OF KEY TERMS"

The New York Times Magazine

MARCH 29, 1970   SECTION 6

Marching for a better earth

Contents—Page 5

2.3
Cliff Humphrey and members of Ecology
Action pushing a planet earth in a stroller,
photographed by Jim Salk for the cover
of the *New York Times Magazine*, March 29,
1970. Courtesy of *The New York Times*.

There was a third group, too, that was neither at the cocktail party nor among the festival-like gathering in the meadow. This group included Jean Baudrillard, the French philosopher and sociologist, and the architect Jean Aubert. They were members of Utopie, a Paris-based collective of thinkers and architects that, between 1966 and 1970, had been engaged in radical leftist critique. In fact, this third contingent comprised thirteen special guests, who would become known collectively at the conference as the French Group, and had been selected by industrial designer Roger Tallon.

Here were three very different tribes, each with its own design principles, conception of what the environment meant in relation to design, and critical methods. To the IDCA board, design was a problem-solving activity in the service of industry—albeit with roots in architecture and the fine arts. In the environmentalists' and students' view, designers needed to claim responsibility for the repercussions of their activities, and to understand design in terms of interconnected systems and natural resources. And the French Group? Well, they objected to both conceptions, perceiving the gathering at Aspen to be a "Disneyland of Environment and Design" and, indeed, the "entire theory of Design and Environment" as constituting a "generalized Utopia ... produced by a capitalist system that assumes the appearance of a second nature, in order to survive and perpetuate itself under the pretext of nature." With such divergent worldviews and reference bases at play, and the prospect of a week in one another's company, an ideological collision of some significance seemed likely.

Sure enough, as the June week wore on, tensions mounted between members of the American liberal design establishment and the eclectic assortment of environmentalists, design and architecture students on the one hand, and the French delegation on the other. The new arrivals to the conference were coming from very different places, both geographically and ideologically. But, in combination, their protests, which became evident during the event, targeted what the dissenters saw as the conference's flimsy grasp of pressing environmental issues, its lack of political engagement, its hubristic belief in design's power to solve social problems, and its outmoded nonparticipatory format.

INTRODUCTION

"Rabble-rousing" versus writing

The critique that materialized at the International Design Conference at Aspen in 1970 epitomized more widespread clashes that took place during the late 1960s and early 1970s between a counterculture and the dominant regime over issues such as the US government's military intervention in Vietnam, the draft, and the civil rights movement. In terms of design discourse, the protests connected with broader challenges to modernist orthodoxies represented by the work of Italian radical architecture collectives such as Superstudio and UFO.

The ways in which design criticism itself was advanced were also under attack. The design establishment, represented by the conference organizers, favored consensus-building as a goal of discussion and a lecture format whereby speakers delivered long, nonvisual, prewritten papers from a raised stage to a seated audience. Dissenters at the conference, interested in participatory formats that could incorporate conflict and agonistic reflection, introduced theatrical performances, games, workshops, and happenings, and confronted the conference organizers directly with a series of resolutions they wanted attendees to vote on. As Ant Farm member Chip Lord remembered it: "Once we arrived we did not have passes to attend all the sessions, so we became rabble-rousers around the edge."[3]

By eschewing the written text in favor of "rabble-rousing," physical actions, and the spectacle of a public vote, the protestors at Aspen disrupted design criticism itself, which, in this period, was usually rendered public in its written form. As such, it was practiced within structured institutional environments where the basic assumptions of design's role in society were generally agreed upon, and points of difference were debated using historical precedents and examples within a common frame of reference. So, although a design critic writing in the 1950s and early 1960s might have been critical, he or she was operating within a reformist tradition rather than a revolutionary one, and his or her criticisms were still contained within the pages of a publication usually paid for, and published, by upholders of establishment values.

For the most part, written design criticism of the period was a one-way communication. Critics could gauge response to their articles only indirectly through letters published in subsequent issues of the magazine; mostly their criticism was uttered into a silent void. As Jean Baudrillard wrote in his 1971 essay "Requiem for the Media," "the entirety of contemporary media architecture" is based on the fact that "it speaks and no response can be made."[4]

2.4
Still from *IDCA '70* showing the Moving Company doing a theatrical improvisation in which they exaggeratedly wiped themselves with toilet roll on stage before leaping into the aisles and attempting to charge five cents to get into Never-Never Land. Dir. Eli Noyes and Claudia Weill, IDCA, 1970. Courtesy of Eli Noyes.

With the criticism at IDCA 1970, the situation was different: The protest that punctured the conference was made up of numerous nonwritten, ephemeral elements, including corridor discussions, Q&A sessions, attire, body language and gestures, theatrical performances, inflatable structures, parties and picnics, objects, and graphic ephemera. These facets were recorded kaleidoscopically in photographs, a film, and audio recordings of presentations and discussions (which include the audience comments that were shouted out). In combination they represent a form of criticism as a spontaneous and performative event, which used countercultural activist tactics as a "style of action."[5]

The protestors were able to confront their targets in person, and could register the effects in real time. The multipronged internal critique of the conference led to a transformation of its content and structure not just in 1971, which saw the most emphatic demonstration of response and change, but also in subsequent conferences at least through the mid-1970s. This makes the events of IDCA 1970 a particularly illuminating case study of a disruption to, and a paradigm shift in, the established practice and role of design criticism in the postwar era.

IDCA '70: the movie

Both the cocktail hobnobbing of the IDCA board members and the countercultural discontent of the attendees at IDCA 1970 were captured in a twenty-minute documentary film of the conference, *IDCA '70*, made by Eli Noyes (the 28-year-old son of Eliot Noyes) and his 24-year-old girlfriend, Claudia Weill.[6] Recent graduates of Harvard and budding filmmakers in New York, Noyes and Weill had been invited by the IDCA board to document the conference.[7] They were given a budget of $5,000, but no brief. Inspired by the cinéma vérité approach to filmmaking, practiced at that time by directors such as the Maysles Brothers, Weill and Eli Noyes had just spent several months living with a black family in Washington, D.C. to produce the documentary *This Is the Home of Mrs. Levant Graham*. Cinéma vérité played a key role in documenting many countercultural movements of the late 1960s. It was characterized by its departure from documentary traditions such as face-the-reporter interviews and voice-over diegetic narration, thus allowing for a potentially more democratic and nonhierarchical version of events to be presented, a method that seemed particularly appropriate for recording the political and social protests of the period.

As East Coasters in their late twenties, Eli Noyes and Weill were not a part of the West Coast student hippy contingent at Aspen. And while Noyes had grown up in the family home in the modernist design enclave of New Canaan, Connecticut, surrounded by such friends and neighbors as Charles and Ray Eames, Alexander Calder, and Philip Johnson, he had chosen a career path that led away from industrial design and, therefore, did not feel that he fitted easily in the world of the Aspen leadership either.[8]

Noyes and Weill used a state-of-the-art Swiss Nagra tape recorder with a shotgun microphone and a newly available Eclair NPR, a French 16mm camera that, with its preloadable magazines, enabled them to speed up the film changes, and minimize interruption to the flow of content. The camera had a crystal-controlled motor and was designed to ride on a filmmaker's shoulder, so that he could move more freely in and around his subjects. Noyes recalls of the camera: "the eyepiece rotated so you could cradle the camera in your lap and look down into the eyepiece even as you filmed something that was horizontally away from you. We wore a battery pack around our waist. It was innovative for its time."[9]

Noyes and Weill seemed to be ideal documentarians, therefore, since they could move freely among the conference's different constituencies, encumbered neither by personal loyalties nor by technology. In reality, what comes across is not so much their neutrality as their shifting sympathies. Through numerous cuts, the filmmakers used the technique of juxtaposition of contrasting scenes to accentuate their view of a conceptual divide between the modernist organizers of the conference and the countercultural contingent. At times Noyes and Weill got caught up in the excitement of the protests, but they also gave airtime to the board members' points of view, ultimately giving them the last word.

Fish frys and kite flying: early years at Aspen

The International Design Conference at Aspen was conceived in 1951 as a forum for designers and businessmen to discuss the shared interests of culture and commerce at a far remove from their everyday concerns. Its founders were Walter P. Paepcke and Egbert Jacobson, president and art director, respectively, of the Chicago-based packaging company the Container Corporation of America (CCA), which was well known for its integrated corporate design. As Jacobson pictured it, a conference that included opinion makers of the American business world "would give the designers a chance to present their case to men ordinarily difficult to reach. For while such men would probably not be tempted to come to hear a speaker like Herbert Read on 'Education through Art' they might be willing to make an effort to hear business peers on the very same subject."[10] This unabashed fusion of high ideals and shrewd pragmatism was not unique to Jacobson; it informed the tenor of many subsequent design conferences at Aspen.

The conference leadership wanted to encourage business executives to apply design cohesively throughout their entire organizations, from letterhead and advertising to truck livery and office design, just as it was implemented at firms like CCA. "Good Design is Good Business" was considered as a title for the first conference and, even though it was rejected in favor of the less blatant "Design as a Function of Management," this remained the IDCA's unofficial motto throughout the 1950s.[11] In a speech to the Yale Alumni of Chicago, excerpted in the advertising brochure for the 1951 conference, Paepcke said: "a Design Department, properly staffed, and given support and wide latitude, can enhance a company's reputation as an alert and progressive business institution within and without its organization, and assist materially in improving its competitive position."[12]

The conference's loftier aim was to imbue businessmen with cultural responsibility and humanist values, and was part of Paepcke's larger mission to promote the arts and culture within American society. Paepcke and his wife Elizabeth had helped develop Aspen from a deserted silver mining town into a winter ski resort and summer cultural festival destination in the late 1940s. In 1949 Paepcke commissioned Finnish architect Eero Saarinen to build a tent for his first cultural event, the Goethe Bicentennial Festival, a twenty-day gathering which attracted such prominent intellectuals and artists as Albert Schweitzer, José Ortega y Gasset, Thornton Wilder, and Arthur Rubinstein, along with more than 2,000 other attendees.

In 1950 Paepcke established the Aspen Institute of Humanistic Studies, an idealistic think-tank with the goal of extending a crusade for the reform of American higher education that University of Chicago president Robert Hutchins and philosopher Mortimer Adler had begun in the 1930s and 1940s.[13] In a 1951 brochure the Institute described itself in the following high-minded terms: "The essence of its humanistic ideal is the affirmation of man's dignity, not simply as a political credo, but through the contemplation of the noblest work of man—in the creation of beauty and the attainment of truth."[14] As historian James Sloan Allen argues, the Institute's version of humanism emphasized the application of reason to scientifically irresolvable questions of principle and value. "Thus 'humanistic studies' meant an analytical way of thinking sharpened by repudiation of the moral relativism associated with empirical science."[15]

The IDCA, formed as an offshoot of the Aspen Institute, with the aim of increasing understanding between business and culture, was timed to run at the end of June each year right before Aspen's summer program of music and cultural discussion, which started at the beginning of July, with the intention that some businessmen would stay on for this too. IDCA promotional brochures of the period used exalted language similar to that of the Institute, referring to design, for example, "in its larger concept as one of the important distinguishing features of our civilization."[16]

Two hundred and fifty designers and their spouses attended the first IDCA, at which top-billed speakers included, on the business side: Stanley Marcus, president of Neiman Marcus; Andrew McNally III of Rand McNally; Harley Earl of General Motors; and Hans Knoll, president of Knoll Associates. Representing design and architecture were: Josef Albers, then a teacher at Yale University; architect Louis Kahn; industrial designers and architects Charles Eames and George Nelson; and graphic artists such as Leo Lionni, Ben Shahn, and Herbert Bayer.

With the exception of Paepcke, the conference leadership came from the design camp, however, and, over the years, they were unable to sustain the participation of business leaders. As the conference evolved, and particularly after Paepcke died in 1960, attempts to improve the dialogue between designers and their clients were abandoned (although the topic was ever-present) and the conference mission broadened to include almost any subject that the leadership believed design touched or was touched by. Scientific philosophers such as Lancelot Law Whyte and Jacob Bronowski, the microbiologist René Dubos, African-American poet Gwendolyn Brooks, and the composer John Cage, for example, were typical of the participants from other professions that began to populate the speaker rosters. And throughout the 1960s the conference was used as a forum to introduce social and behavioral sciences to architectural and design discourse.

While the scope of the conference expanded, and the theme changed from year to year, the format remained the same. Speakers addressed conferees from a raised stage in Saarinen's large, tented auditorium (which was replaced in 1965 with a new one by Herbert Bayer). Even though speakers were asked to prepare 45-minute papers, they were rarely that short, and there was little opportunity for improvisation since speakers' presentations were printed and circulated ahead of time, and conferees were "advised to read each speaker's paper in advance of the session."[17] Daytime lectures in the tent were delivered without images; slide presentations were scheduled in the evenings when it was dark enough for projections.

2.5
Leaflet for the 1969 IDCA showing the tent
designed by Herbert Bayer in 1965.

Aspen was notoriously difficult to reach. In the early years of the conference, telephone and telegraph service was unreliable and there was no radio or television. Several speakers recounted the experience of traveling to the conference and the challenging last leg of the journey from Denver to Aspen, either by car on poorly finished winding roads or in a small twin-prop plane. But the remoteness of the conference location was, of course, part of its appeal. Paepcke had always hoped that attendees would return home renewed in body and spirit, as well as in mind. The pace was leisurely, with presentations spread out over a week and interspersed with long lunches and rambles in the surrounding mountains. An annual favorite was the Fish Fry, an al fresco lunch by the river. A typical outdoors afternoon event was billed as "A discussion and demonstration of international kites, led by Charles Eames and Michael Farr."[18]

2.6
Photograph of Charles Eames and guests at
IDCA cocktail party, by Ferenc Berko.
Courtesy of The Getty Research Institute.

"CONFLICTING DEFINITIONS OF KEY TERMS"

In the evenings there were cocktail parties by the pool at the Hotel Jerome. The brochure for the 1961 conference dispensed the following advice on attire: "Sportswear is the norm for the daytime, and evening dining is only a shade more formal. At the Monday night IDCA cocktail party at the Jerome pool, a little black dress and mosquito repellent will do for the ladies, and a plaid coat, tie and Bermudas for the men."[19]

The design historian Nikolaus Pevsner attended the conference in 1953, and on his return shared his impressions with British listeners on a BBC radio broadcast. Pevsner was fascinated by the casual attire of the attendees, their "coloured printed" and "wildly patterned" shirts, in which he went so far as to locate the source of America's advanced progress in modern industrial design: "I am, as a matter of fact, quite ready to appreciate these shirts intellectually, and if that daring, that naive trust in novelty were not part of the American character, modern design of the best quality would not have made such spectacular progress in the last ten years—along, of course, with modern vile design."[20]

Among this extravagantly shirted and collegial group of IDCA board members, there was a shared belief in what constituted good design, and, where opinions differed on points of detail, a shared belief in the worth of debating an issue toward the goal of mutual understanding. This desire to forge consensus derived from the conference's origin as an offshoot of the Aspen Institute. Even in 1970, many of the conference's organizers still espoused the humanist values advocated by the Institute and by liberal social theorists of the early 1950s such as David Riesman and Erving Goffman.

Throughout this period, the IDCA was the only design conference of its kind, and a key event on the international design agenda. Thanks to the dissemination of speakers' papers and extensive press coverage—whole issues of design magazines such as *Print* and *Communication Arts* were devoted to it—the conference's influence extended well beyond the 1,000 or so attendees it attracted each year. As Reyner Banham observed, the IDCA was "the most heavily reported design conference on the calendar, outranking even the Triennale di Milano, let alone the biennial congresses of the International Council of Societies of Industrial Design."[21]

In 1954, the IDCA was formalized as a not-for-profit corporation, headquartered in Chicago, and administered by an executive committee that elected their own president. The organization was funded by membership dues, conference fees, and industrial sponsorship. By 1970, therefore, what had started as an experimental meeting had evolved into a robust institution. As the American cultural climate underwent dramatic change toward the end of the 1960s, and a younger generation of more politicized designers emerged whose practices incorporated critique, the IDCA, representing the status quo among the higher echelons of industrial design, graphic design, and architecture, was ripe for attack.

The format problem
Many of the attendees at IDCA 1970, captured on film by Eli Noyes and Claudia Weill, were aggrieved by the conference format, which they saw as "outmoded." "It's curious to me

2.7
Spread of photographs of IDCA speakers and
scenes, 1951–1974, in Reyner Banham,
The Aspen Papers (New York: Praeger, 1974).
Courtesy of Phaidon Press.

2.8
Spread of photographs of IDCA speakers and
scenes, 1951–1974, in Reyner Banham,
The Aspen Papers (New York: Praeger, 1974).
Courtesy of Phaidon Press.

"CONFLICTING DEFINITIONS OF KEY TERMS"

that change is so long in coming to this design conference," a bearded youth told the film-makers. "It's one speaker and 1,000 people glued to their seats by regulation, or boredom, or both."[22] Another attendee remarked: "Nobody wants to sit passively and listen anymore."[23]

The format itself became symbolic of the inadequacies of the prevailing regime, and of the potential for a new vision of participatory information exchange. The one-way transmission of information from a designated expert on a raised stage to a seated audience was seen as anachronistic in this period of experimentation with new modes of communication. Many schools were being reinvented in the form of free universities or anti-universities. Roberta Elzey's contemporary account of the Anti-University in London explains the nonhierarchical principles of the movement: "Anti-University classes were totally different from those at academic universities, as were the roles of 'teacher and student.' These were fluid, with students becoming teachers, and teachers attending one another's classes. About half those in Francis Huxley's course on Dragons were Anti-University teachers at other times. There was one lounge, used by all: no sacrosanct staff lounge or common room."[24]

At campuses across the US, too, particularly in California, new educational configurations were being tested. The California Institute of the Arts (CalArts), for example, was established in 1970 and, through an educational program of independent study and non-hierarchical teaching relationships with artists such as Allan Kaprow and Nam June Paik, hoped to provide what the dean, Robert W. Corrigan, characterized as a radically different prototype for training the artist of the future. Ant Farm, who visited numerous California schools during the academic year 1969–1970, described their teaching work—"lectures, ecology events, environmental alternative displays, or art"—as "response information exchanges."[25] Yet, even though the topic of format often came up in IDCA board meetings throughout the 1960s, conference chairmen inevitably returned to the same lecture setup dictated to them by the interior architecture of the tent.[26]

2.9
Still from *IDCA '70* showing seated audience.
Dir. Eli Noyes and Claudia Weill, IDCA, 1970.
Courtesy of Eli Noyes.

The student problem

Students presented the IDCA leadership with another problem more pressing and perennial than the format issue. In the conference's early years, design and architecture students had attended in small numbers, gaining free admission in return for their labor. They escorted speakers between the airport, the hotel, and the main tent, helped with audiovisual equipment, ran errands, and cleaned up.[27] As they began to attend in greater numbers, and were asked to pay a fee, they made more demands of the conference, such as involvement in the planning and travel grants, and in 1968 a group of them set up their own Student Commission to organize such demands.

The twice-yearly meetings of the IDCA board devoted more and more time to the discussion of students. The board members doubted the students' "seriousness" and were unsure about what kind of contributions they could actually make. Board members at the post-IDCA 1969 meeting noted that students "seemed to be in about the same mood as in the previous year, lacking direction, being considerably confused, and yet groping for some additional identification."[28] None of the board members mentioned the student protests that had filled the streets of Paris the previous summer, but one of them, the urban planner Julian Beinart, observed that "the student problem had to be handled in a most flexible manner, since it is impossible to predict much about them or their attitudes."[29]

The lead-up to the 1970 conference saw an intensification of the student attendees' dissatisfaction with their peripheral role. Minutes of the planning meeting prior to the conference show that board members still assumed that the students' gripes could be appeased by giving them more responsibility and "a desk somewhere."[30] The students, who in 1970 represented a larger proportion of the conference community than ever before, had other plans.

The planning of an "anti-conference"

As Sim Van der Ryn remembered it, in the month preceding the conference, "the Aspen board got word that a number of long hairs and radical edge groups planned to show up and stir up the stodgy elitist establishment Aspen Design Conference."[31] Van der Ryn was asked to step in and manage these students and environmental action groups because, as a professor at the University of California, he could be considered to be someone within the "establishment" who also had connections and sympathies with radical groups. In May 1969, student protestors who attempted to claim an empty lot belonging to the University of California at Berkeley for a park and site for demonstrations were fired on with buckshot by police, under orders from Governor Reagan, who saw the creation of the park as a leftist challenge to the property rights of the university. Van der Ryn had been the university negotiator in this incident, when, as he remembered it, "Ronald Reagan called out troops and helicopters to spray poison gas."[32]

The students and activist groups at IDCA 1970 had been invited to submit a proposal to create something at the conference, which would be eligible for funding from the Graham

Foundation. The previous year, Northern Illinois University students had used their funding to create a sculpture of junked cars, toilets, sinks, and old tires, sprayed white, intended to embody the current state of contemporary design.

When the environmental groups' proposal for the 1970 conference was received, however, it was not for a sculpture (a material form that the conference leadership understood); rather, they wanted to use the funds to bring thirty-five people from their organizations to Aspen in a chartered bus, giving small theatrical performances along the way for several weeks. They proposed to set up inflatable structures in Aspen, in which to hold meetings and exhibitions, present performances, and create a series of events that would, it seemed to Eliot Noyes, "be in conflict with the Conference itself, almost as a counter-conference, or an anti-conference."[33] The premeditated nature of the ensuing protest, revealed in this correspondence, suggests the revolutionary nature of its purpose. According to cultural critic John Berger, writing in 1968, demonstrations of the period, in their very "artificiality" and "separation from ordinary life," were means of "rehearsing revolutionary awareness."[34]

STUDENT HANDBOOK   INTERNATIONAL DESIGN CONFERENCE AT ASPEN, 1970

2.10
Cover of *Student Handbook*, an unofficial publication printed on newsprint and distributed at IDCA 1970, showing a sculpture of junked cars and appliances, painted white and assembled in Aspen by students from Northern Illinois University under the supervision of their professor, Don Strel, in 1969. Courtesy of The Getty Research Institute.

The valuable "artificiality" of the Aspen protests, in Berger's term, was compounded not only by the theatrical nature of their enactment but also by the costumes the protestors wore. No small part of the tensions between the authorities and the unruly student attendees derived from their physical appearance. As Reyner Banham observed: "Once a distinctive student culture began to emerge, taking neither [professionalism and professional status] seriously nor for granted, and began to replace the deferential boy-scoutism of students at earlier Aspens, there began to be some sense of strain about many human aspects of the conference—not least its relations with the worthy burghers of the business community in Aspen itself, who had a well-nourished paranoia about long hair, bare feet, and all the rest of it."[35]

Most provocative to the Aspen community, however, was the students' intention to sleep outside in inflatable structures, rather than in the guest houses where most attendees stayed. The Aspen Institute, which lent the Aspen Meadows location to the IDCA each year, notified the IDCA board that no structures might be built on Institute grounds around the tent if there was any chance that students would spend the night in them.

The threat to the establishment in the notion of students sleeping in tents had also been at the core of the disturbances at the 1968 Democratic National Convention in Chicago.[36] Protestors, including members of the Youth International Party—better known as Yippies, and led by Abbie Hoffman and Jerry Rubin—converged on Chicago to support Eugene McCarthy and his antiwar platform against Hubert Humphrey. The protests, which took the form of satirical street theater—or put-ons—and the violent response by the Chicago police force, were captured by multiple television news channels, and chronicled by journalists including Norman Mailer and Hunter S. Thompson. The Yippies had planned a weeklong schedule of events under the heading "A Festival of Life," which included "a workshop in drug problems, underground communications, how to live free, guerrilla theater, self defense, draft resistance, communes, etc."[37] The ensuing clashes between the Chicago police force and the protestors lasted for eight days. As journalist Mark Kurlansky recounted, the Yippies' planned program of events "was in conflict with the Chicago police because it was based on the premise that everyone would sleep in Lincoln Park, an idea ruled out by the city."[38]

Two years later in Aspen, disregarding the conference organizers' stipulations that visitors should not bring their own tents, Ant Farm promptly erected *Spare Tire Inflatable*, a tubelike inflatable, twelve feet in diameter, which they had created earlier that year.[39] Power for the air pumps was supplied by their Media Van, in which they had traveled to the conference. When asked by the *IDCA '70* filmmakers why they were at the conference, a founding member of Ant Farm responded: "We ripped off $2,000. We're here on vacation like everyone else" (referring to the grant given by the IDCA board to the five invited environmental action groups to enable them to attend the conference).[40] Ant Farm member Hudson Marquez, captured on the film sporting a bushy beard, beads, and dark sunglasses, expanded: "We wanted to go to Boston to shut down the AIA conference but we didn't have money to get there. So we pushed buttons and pulled levers and threatened to have thousands of hippies show up at Aspen. We said we were going to put an ad in the underground newspapers in Berkeley advertising free food and hanging out with Aquarian age architects and all that bullshit. I guess they bought it."[41]

Marquez's comment suggests that the protestors planned more than discourse. As part of a growing critique against corporate modernism and rationalist approaches toward design, and possibly inspired by the well-publicized attempt to "close down" the city of Chicago in 1968, students and activists occupied other design conferences of the period. The American Institute of Architects' (AIA) annual conference in Boston (which was running concurrently with IDCA 1970) was subject to a revolt in which students, led by Taylor Culver, took over the podium from the AIA president, Rex Whitaker Allen.[42] Similarly, Utopie member Hubert Tonka has recalled going to the "Utopia or Revolution" conference organized by the architecture department at Turin Polytechnic in April 1969: "We held the whole conference hostage for several hours with a leftist group called the Vikings. The cops showed up with submachine guns, etc. Oh yes, 'Utopia or Revolution,' that was a bad scene."[43] Also, in May 1968 radical demonstrators in Milan had protested against the elitist organization of the Milan Triennale, its aestheticization of the Paris student protests, and its reformist approach to that year's theme of "World Population Explosion" (A translation of the Italian phrase *Il Grande Numero*).[44] Students erected banners that read "The Triennale is Not Paris—Merde to the Falsifiers," managed to close the Triennale down only hours after it had opened, and provoked the resignation of the event's executive committee.[45] By 1970, therefore, the design event had already been identified as a public stage upon which to resist the design establishment.

2.11
Still from *IDCA '70* showing members of Ant Farm taking a Polaroid photograph of the filmmakers. Dir. Eli Noyes and Claudia Weill, IDCA, 1970. Courtesy of Eli Noyes.

"Environment by Design": differing definitions

While the traditional format of the conference invited attack, IDCA 1970's theme rendered it still more vulnerable. What transpired at IDCA 1970 reveals that two very different definitions of the concept of "environment" were being wielded in design discourse and beyond, and highlights the conceptual fault line along which the conference would ultimately split.

For the most part, the IDCA leadership considered "environment" to be simply the background against which their designed images, products, buildings, and urban plans would operate. When IDCA had devoted another conference to the topic in 1962, chaired by Ralph Eckerstrom, CCA's director of design, advertising, and public relations, they had portrayed "environment" as a "physical setting" which could expand along a spectrum of scale: "a room, a house, a city, a countryside, a nation, the world—the universe."[46] A consideration of the environment, for the 1962 conference organizers, was closely tied to a consideration of aesthetics. The "critical problem" in environment, to them, was the difficulty of isolating technological advances and good design from the polluting presence of mass culture: "Wider windows of distortion-free glass for better transmission of uglier vistas; higher fidelity for clearer reception of cacophony. ... Mass production for endless repetition of the meretricious." This discussion of the environment as an arena for one's work, often subject to aesthetic assault by unchecked development, was continued at IDCA 1970 by speakers Stewart Udall, James Lash, Reyner Banham, and Peter Hall, who spoke of urban decay, ghettos, and the possibility of renewal through New Towns.

The chair of the 1970 conference was William Houseman, editor and publisher of *Environment Monthly*, president of the Environment League, and a charter board member of the Institute of Environmental Design in Washington, D.C. His biographical statement in the conference brochure emphasized his generalized interpretation of environment, indicating that "his interests in the subject range from the Aviation environment to the role of color and design in the everyday lives of people."[47]

In his opening remarks, Houseman further confirmed that his interpretation of the concept of environment as the backdrop for design, not a political issue, was firmly aligned with that of the IDCA board. Quoting "our good friend" George Nelson, he said: "When you walk down any street in any town you will find endless objects that are objects of design ... the man hole covers ... mailboxes, screen doors ... they have all got design."[48] For many of the conference organizers, then, environment was, quite simply, the urban context for their work.

For the ecology groups, on the other hand, "environment" meant a pressing political issue—the need to protect the earth's natural resources from further destruction at the hands of the dominant political and economic interests. As Ecology Action founder Cliff Humphrey said in his main-stage lecture, "What we are talking about, then, is manifesting by design a survival gap—a survival gap between the people on this planet and the ability of the life support system to support these people."[49]

Clifford C. Humphrey
writer, lecturer, teacher
Clifford Humphrey is Director of the Ecology Action Educational Institute, a non-profit Corporation. He is also co-founder of Ecology Action, an organization with over 150 groups throughout the country at present. His major concerns are the cultural implications of the ecological perspective and the development of a system of ethics for human behavior that will insure our survival. His book, "What's Ecology?" is a widely used high school text. He is the author of many essays, position papers and statements including, "Declaration of Interdependence". He has also written texts for two large exhibits: The Environment of Man and The Survival Crisis, which are displayed on college campuses, at museums, shopping centers and public parks. He is a consultant at Berkeley High School, Berkeley, Calif., on curriculum and has taught the course, "Ecological Dynamics of Social Change", at the University of California Extension in Berkeley.

2.12
Page from IDCA 1970 speaker biographies book-
let showing portrait photograph and biography
of Clifford Humphrey, founder of Ecology Action.
Courtesy of The Getty Research Institute.

Humphrey made use of an array of visual props on stage, including a pile of garbage gathered during the conference and an image of the earth seen from space (reproduced from the cover of the Fall 1969 *Whole Earth Catalog*) to enact a kind of three-dimensional diagram, demonstrating the urgency of the impending environmental crisis, which he framed in terms of species survival. "If an item is made to be wasted, to be dumped on a dump, then don't make it!" Humphrey proclaimed, to much applause.[50] "You know, if our youth can say 'Hell, no!' to the draft, then I think that a few of you have to learn to say 'Hell, no' to some salesmen and to some developers."[51]

An unofficial *Student Handbook* created for the 1970 conference reported on students' responses to the previous year's conference, and included articles on issues of contemporary interest such as: a *Science* magazine article on the historical roots of the ecological crisis; World Game, a simulation tool for visualizing "spaceship earth" (developed by Mark Victor Hansen and inspired by Buckminster Fuller's Dymaxion sky-ocean map); and yoga breathing. In an introductory sally to the students, the editors of the *Handbook* enumerated what they thought would be the important aspects of the conference, such as which speakers would be worth their attention (all the speakers mentioned were the special guests of Van der Ryn), concluding with a nihilistic amendment to the official conference prose: "According to the official litter bag, we are here to ponder what is worth keeping, what is worth restoring, and what is worth building. (May I add, 'What is worth destroying?')"[52]

In IDCA thinking, the environment could be improved through thoughtful design. From the perspective of a new generation of designers and their environmentalist mentors, the design system (supported by capitalist interests) was an integral part of the environmental problem, and should be resisted and ultimately rejected.

Off-stage activity: new formats tested

The ecology groups initiated numerous interventions during the week of the conference, with varying degrees of success. Among them was an impromptu "Favorite Foods Picnic" on the grass outside the tent.[53] It was Van der Ryn, rather than program chairman Houseman, who invited the ecology groups to participate in the conference, and Houseman's cynical view of their interventions is evident in his flippant "A Program Chairman's Diary of Sorts," included in the post-conference publication. Under the heading "Monday Noon," for example, he satirized the groups' attempt to create, and then clear up, an organic picnic:

> Precedent! For the first time ever, an impromptu Favorite Foods Picnic on the grass outside the tent. The young and otherwise decimated the local shopkeepers' shelves in frantic quest for favorite foods. Mostly salami. Enough for the Bulgarian cavalry. Ecological havoc! Cliff Humphrey officiated at the burial of the picnic's organic residue. But what of the nondegradables? Under the cover of darkness, Aspen's anthropomorphological dogs scattered paper plates and Reynolds Wrap across the greensward. A regular Les Levine sculpture.[54]

The film of the conference, *IDCA '70*, documents an unscheduled session in which the attendees were instructed to stand up and pass their name badge to the next person, and so on, and then embark on a process of relocating themselves. This rather crude attempt at encouraging audience interactivity was instigated by "some of the young people from California," as artist Les Levine described them—namely: Chip Chappell, a teacher at Oakwood School; Tony Cohan, a writer from Los Angeles; and Mike Doyle, leader of the Environmental Workshop and an architect at Lawrence Halprin & Associates.[55] Doyle would go on to become a strategic planner, change consultant, and coach for corporate and nonprofit organization leaders, and to coauthor a best-selling book titled *How to Make Meetings Work* and training films such as *Meetings, Isn't There a Better Way?*[56]

Levine was a special guest of the conference, and wrote a report for the *Aspen Times*. He saw the spontaneous name card exchange as an opportunity to pull out his "Merry Cambodia" and "Happy New War" cards, which he had printed in ornate type. While attendees exchanged cards, talked to one another, and made handmade signs in the search for their name badges, Chappell, Cohan, and Doyle paced about on the stage with hand-held microphones, rationalizing the exercise as a demonstration of the attendees' interdependence as part of an "ecological chain."[57] Cliff Humphrey's militant manifesto, "The Unanimous Declaration of Interdependence," in circulation at the conference, was a neatly wrought subversion of Thomas Jefferson's Declaration of Independence. It declared that "all species are interdependent" and that "whenever any behavior by members of one species becomes destructive to these principles, it is the function of other members of that species to alter or abolish such behavior and to re-establish the theme of interdependence with all life."[58]

2.13
Still from *IDCA '70* showing Chip Chappell on stage during an unscheduled session in which the attendees were instructed to stand up and each pass their name badge to the next person, and so on, then embark on a process of relocating themselves. Dir. Eli Noyes and Claudia Weill, IDCA, 1970. Courtesy of Eli Noyes.

"Conflicting definition of key terms"[59]

In between the speaker presentations on the main stage, attendees gathered in small discussion groups in the Aspen Institute seminar rooms. The *IDCA '70* film shows that IDCA board members made numerous attempts to engage attendees in conversation, even sitting down with them on the floor, but it was clear that the middle-aged modernists and the young environmentalists had great difficulty communicating with one another. Not only did they *look* different, they didn't even share the same basic vocabulary.

S. I. Hayakawa, a linguist who specialized in semantics and would go on to be a US senator, gave a paper at IDCA 1956, which was reprinted and circulated at several subsequent conferences. In "How to Attend a Conference," Hayakawa articulated the gentlemanly code of conduct required from both speaker and listener at an IDCA conference in order to reach consensus. He portrayed the conference as a "situation created specially for the purposes of communication" in which ideas are exchanged and personal viewpoints are enriched "through the challenge provided by the views of others." Discussion is stalemated, wrote Hayakawa, by "terminological tangle," or the "conflicting definitions of key terms."[60]

As if in illustration of this predicament, the *IDCA '70* film includes a particularly heated conversation between some board members, including Saul Bass and Eliot Noyes, and members of the Moving Company theater troupe, one of whom had to explain the then-new term "hype" to a confused Noyes. Subsequently the conversation between a crisp-looking man and the long-haired leader of the Moving Company broke down completely. They leaned in and jabbed their index fingers at one another, as they became visibly frustrated with their inability to find any shared points of reference:

Man: So, you're saying that I have to understand what you're telling me today? I don't understand it.

Moving Company member: We were saying that everything is a rip off. Everyone is stealing. ... The entire civilization is based on the wrong premises. Dig that. We are living in the wrong reality.

2.14
Still from *IDCA '70* showing heated discussion between members of the Moving Company and Eliot Noyes at IDCA 1970. Dir. Eli Noyes and Claudia Weill, IDCA, 1970. Courtesy of Eli Noyes.

Man: Tell me what the right civilization is.

Moving Company member: I can't talk to you if you say that, because you're already saying that you're alienated.[61]

This concept of "alienation," codified most prominently by German-American philosopher Herbert Marcuse, had been key to the student protests in Paris of 1968, and by 1970, through its dissemination in the underground press, had clearly become part of the lexicon of those adopting alternative lifestyles in California. Marcuse's book *One-Dimensional Man* was published in the US in 1964, and popular interpretations of his thinking, such as Paul Goodman's *Growing Up Absurd*, were both listed on the IDCA 1971 reading list.[62]

In another corridor conversation captured by the film, Bass joined a group of students seated on the floor and asked them: "Why do we have to assess capitalism? We're just trying to stage a design conference." A young, intense-looking individual attempted to explain: "Unless you actually live the lifestyle, it's just bullshit."[63] Bass was clearly upset that his attempts to understand these unfamiliar beliefs were rebuffed so emphatically. In the board meeting after the conference, he reflected: "If I walk away from this I will feel defeated as a person. ... This time the design problem is ourselves. That's why I'm so shook up about this whole thing."[64]

With a theme as broad as "environment" under discussion, it is not surprising that multiple definitions were being used by IDCA 1970's different constituents. The severity of the breakdown in communication, however, was new to a conference that prided itself on debating divergent opinions to the point of consensus.

The closing session: the French Group's statement and the students' resolutions

Tensions mounted throughout the week, reaching a crescendo in the closing session on Friday morning. This session centered on voting for a series of resolutions formulated by the protestors that criticized the intellectual and moral limitations of the conference content, the conference as a designed entity, and the design profession itself.

Reyner Banham, who had been a speaker at the conference several times since 1963 and had organized the 1968 edition, titled "Dialogues: Europe/America," was the chair of this closing session. In a letter written later that evening to his wife, in which he said he was feeling "psychologically bruised from the events of this morning," Banham explained that it was actually his idea to turn the final session into a soapbox for the disgruntled attendees. This suggests that he was just as eager as the IDCA board members to resolve the dispute: "This has been too fundamentally disorganised a conference to sum up—intellectually disorganised, that is—Bill Houseman really hadn't got the programme together enough for it to gel, and the kinds of people he had invited (from ex Secretaries of State to the Ant Farm Conspiracy) were a guaranteed communications failure. So I proposed we use the morning for second thoughts, statements, and the like."[65]

Banham's self-imposed challenge of consensus building was made particularly tough by the fact that the goals of the groups who converged in this session—from Stephen Frazier's group of fifteen Black and Mexican-American industrial design students from Chicago to the seemingly arbitrarily selected group of French participants—were so ideologically heterogeneous.

The French Group's contribution to the conference was a statement written by Baudrillard that explained the group's refusal to participate in the regular conference proceedings.[66] In their view, essential matters concerning the social and political status of design were not being addressed. "In these circumstances," the statement began, "any participation could not but reinforce the ambiguity and the complicity of silence which hangs over this meeting."[67]

Baudrillard's text, read aloud at the closing session by the geographer André Fischer, openly dismissed the conference's theme of "Environment by Design." It also rejected the more widespread interest in environmental issues, as an opiate concocted by the capitalist system to unify a "disintegrating society."[68] Baudrillard posited that both the conference theme and the wider crusade currently preoccupying the nation simply diverted attention and energy toward "a boy-scout idealism with a naive euphoria in a hygienic nature," and away from the real social and political problems of the day such as class discrimination and neoimperialistic conflicts. The new focus on pollution, Baudrillard pointed out, was not merely about protecting flora and fauna, but about the establishment seeking to protect itself from the polluting influence of communism, immigration, and disorder.[69]

During an informal debate with Cora Walker, the only black speaker on the conference program, with the audience seated cross-legged on the floor of a seminar room, Jivan Tabibian, a Lebanese-born political scientist who had been educated in French schools and later became ambassador to Armenia, also pointed to the liberal design establishment's "utopian" belief that increased understanding would enable change. He remarked: "what I call the great fallacy of the men of good will totally overlooks the concrete reality of vested interest, of institutional power. Those things don't change just because people understand."[70]

2.15
Still from *IDCA '70* showing group discussion at IDCA 1970, with participants seated on the floor. Dir. Eli Noyes and Claudia Weill, IDCA, 1970. Courtesy of Eli Noyes.

"CONFLICTING DEFINITIONS OF KEY TERMS"

Far from espousing environmentalism, then, Baudrillard contended that it was a ruse of government to maintain the very economy that threatened the environment. He may have been referring to Earth Day, first celebrated in April 1970 and founded by Senator Gaylord Nelson, a liberal democrat from Wisconsin. As historian Felicity Scott notes, it had "set out to repackage environmental concerns for the general public by decoupling questions of ecology from more radical elements and bringing the movement into alignment with those in Congress pursuing environmental regulations. With re-election campaigns in the works, a cynical 'war on pollution' had been added to those already launched through the media on poverty and hunger."[71] Baudrillard identified an insidious "therapeutic mythology" at work, which framed society as being ill, in order that a cure might be offered. He castigated designers, "who are acting like medicine-men towards this ill society," for their complicity in such myth-making, in this semantic slippage between the realms of military defense, the environment, and society.[72]

This statement, however powerful, did not have much impact at the conference. French journalist Gilles de Bure reported that "the text was greeted with polite applause. Neither interrupted, nor discussed, it provoked a reaction of surprise at the most elementary level. ... One may wonder if, in the end, the text by Jean Baudrillard had hit home at all, other than with the French group, which had accepted it even before he wrote it?"[73]

Reflecting on what occurred at Aspen in 1970, Baudrillard identified yet another communicative rupture between his own Utopie collective and the Californian environmental activists—one based on national identity: "This 'counter-culture' was foreign to us. We were very 'French,' therefore very 'metaphysical,' a French metaphysics of revolt, of insubordination, while the counter-culture that expressed itself in Aspen was largely American." When he tried to bring back to France something of the "vigor" of the American movement, he found there was a translation barrier: "There was no way to metabolize this contribution in a French context dominated by the 'politico-careerist' New Left."[74]

Despite Baudrillard's retrospective enthusiasm for the Aspen "moment," he found the physical setting of the conference to be fundamentally at odds with the seriousness of the issue at stake, referring to Aspen as "the Disneyland of environment and design," and drawing attention to the fact that "we are speaking ... about apocalypse in a magic ambiance."[75] Cora Walker had also highlighted the surreally removed location of the conference, telling the crowd: "When asked if I'd ever been to Aspen before, I had to respond that I'd never even *heard* of Aspen before."[76] The high-altitude resort of Aspen, which had once been seen as the ideal setting for designers to gain critical distance from their practices, was now being criticized for its physical and symbolic remoteness from the social problems they should be engaging with.

As moderator, Banham was able to control the final session to only a limited extent. He contrived to hold back what he thought would be the "most explosive items" until after the coffee break. The first part of the morning, he told his wife in the four-page letter he wrote that night, went quietly: The French Group's statement he considered "tough, but gentlemanly," and the "Black Statement," presented by Stephen Frazier, he saw as "routine stuff ... just the usual threats 'we're together and we're *here*, baby'—though effective enough when addressed to an uptight white liberal audience."[77] In fact, Stephen Frazier, a black industrial designer who brought a group of fifteen black and Mexican-American students from Chicago to Aspen, was given a standing ovation for an impromptu speech that drew attention to the symbolic nature of the black students' presence at the all-white design conference.

The students' resolutions, read aloud after the coffee break by Michael Doyle, shared some of the same goals as the French Group's statement. The resolutions called for, among other things, the withdrawal of troops from Southeast Asia and an end to the draft, the legalization of abortion, the restoration of land to Native American Indians, and the end of government persecution of "Blacks, Mexican-Americans, longhairs, homosexuals, and women."[78] But it was the final point of the document that was the most contentious: it asked that designers attending the conference "refuse to create structures, advertisements, products, and develop ideas whose primary purpose is to sell materials for the sole purpose of creating profit," stating that "This attitude is a destructive force in our society."[79] Striking at the core of the design profession, as it was represented by the conference board, this resolution also pointed to the contradiction in the conference's environmental theme being discussed—and, indeed, sponsored—by those deeply implicated, through their day-to-day transactions, in harming the environment.

Stewart Udall, who had been Secretary of the Interior from 1961 to 1969, observed in his keynote speech that Walter Paepcke "would be amused in 1970, if he were here, to realize that the container industry is in trouble, and on the defensive with the environment movement."[80] In 1970, possibly under pressure from a mounting environmental movement, the firm that Paepcke had founded, Container Corporation of America, sponsored a contest to create a design that would symbolize the recycling process, and would be used to identify packaging made from recycled and recyclable fibers. Gary Anderson, a graduate student at the University of Southern California in Los Angeles, won the contest, which was judged during IDCA 1970, with his design based on the Möbius strip, and was awarded a tuition scholarship of $2,500.

Very few of the IDCA board members and speakers at the 1970 conference could claim to work for companies whose main goal was *not* to "sell materials for the sole purpose of creating profit," and even fewer worked for companies with environmentally responsible practices. The corporate contributors included Alcoa, Coca-Cola Company, Ford Motor Company, IBM, and Mobil Oil, all well known for their resource-heavy manufacturing and distribution processes.[81]

After reading the resolutions aloud, Doyle hectored the conference attendees into voting on whether or not to adopt them. Banham noted: "It immediately became clear that the conference was liable to polarize into irreconcilable factions and split as the tensions of the week came to the surface."[82] It was apparent to him that, even though Noyes and most of the board were "clearly frightened and didn't want it voted," what he called "the Berkeley/Ant Farm/Mad Environmentalist coalition" wanted to commit the conference through a vote.[83] He suggested that it could be rephrased as a petition, "if only as a way of getting the pressure off honest folks who were frightened of looking conspicuous in the ensuing mob scenes if they didn't vote."[84] He deliberately kept the debate going on this point by calling on the loquacious Jivan Tabibian, and "picking up every point from the floor, in order to give frightened souls a chance to slip out quietly (they didn't of course; they went out conspicuously later, and got shouted at and threatened)."[85]

2.16
Photograph of conference attendees and speakers, including Reyner Banham and Jivan Tabibian, at IDCA 1968, by Ferenc Berko. Courtesy of The Getty Research Institute.

Doyle denounced the idea of a petition as a "cop out," but Banham did manage to persuade the assembly that the resolution should be voted clause by clause, not as a package, in an effort to overturn the final anticorporate design proposition.[86] Banham's personal frustration with the whole event is evident in a parenthetical aside in the letter to his wife: "(I was doing the whole show single-handed without a whisper of help from Houseman or the Board. In fact, there were a couple of moments during the shouting when I was sorely tempted to pull the plug on the whole operation and leave the Board with the shambles I felt—at that time—they deserved.)"[87] At the end of the session, by Banham's reckoning, only half the conferees remained. "I shall not soon forget the hostility vibes that were coming up from the floor," he wrote, "nor how uptight the students could get the moment they thought they weren't getting their own way."[88]

As moderator of the closing session, the 48-year-old Banham found himself in an awkward position: as an educator and sympathizer with student sit-ins that had taken place in London in the last two years, he wanted to give the students and environmentalists airtime. Less than a decade before this, students at the Bartlett School of Architecture, University College London, had invited Banham to give lectures for their own alternative course, which they were running concurrently with the official degree program. By 1970, however, he was an officially appointed professor at the college's newly formed School of Environmental Studies. Furthermore, as an advisor to the IDCA board, a prior conference chairman, the editor of *The Aspen Papers*, and a close friend of Noyes, he also felt loyalty toward the conference organizers against whom the protests were directed. Ultimately, Banham adhered to the consensus-building tendency that had characterized IDCA to date. By contrast, the writer Tony Cohan, who traveled to the conference with the California environmentalists, advocated "dissensus," calling for a new conference format in which "the thrust would have been away from language and toward action encounter, away from fruitless attempts at consensus and toward forms that incorporate conflict."[89]

Only the year before, at the 1969 conference, titled "The Rest of Our Lives"—and as if he were speaking directly to the following year's attendees—George Nelson had given a speech in which he warned of the self-perpetuating nature of establishments, despite the efforts of the hippies to overthrow them: "But let us rejoice prematurely at the impending doom hovering over the establishments, for the blanket-carrying party members of the young (I'm referring to the party founded by Linus, not Marx, Lenin and Engels) and the bearded, barefoot conformists are presently going to set up new establishments no better or worse than the old ones."[90]

The question of how to engage with, and how to resist, the liberal establishment preoccupied the earnest and impassioned students at the Aspen conference, just as it did students more generally in the late 1960s and early 1970s. It was clear, however, that new forms of resistance were necessary; mere criticism as it was conventionally practiced in a written form was no longer suited to the task. During a meeting arranged by the student attendees of IDCA 1969, to which they had invited some of the speakers (including Nelson), discussion had turned to the widely publicized attempt to create a public park in Berkeley, and whether or not to work with the establishment, to become a part of it, try to destroy it, or to create a new establishment. The report of the meeting records: "Finally one student in anger said, 'You can't write a letter to a vending machine; you have to kick it!' Again there was applause."[91]

After the storm: the IDCA board meeting

On Saturday, June 20, 1970, the morning after the stormy closing session, the IDCA board members gathered in an Aspen Institute seminar room, along with Fred Noyes, Eliot Noyes's other son, who was invited at the behest of some board members to be an interpreter for the foreign-seeming student contingent. Asked if he could describe what the students would like the conference to become, Fred Noyes resisted the idea that the students could be considered a unified body with one easily communicated point of view. By the time of the meeting, however, the directors had convinced themselves that this was a "them and us" situation. Henry Wolf said: "Unless we design a form where all this energy can be used there will be a takeover."[92]

The discussion returned repeatedly to the failures of the conference format. Toward the end of the three-and-a-half-hour meeting, Noyes, who stated that the conference had left him "battered, bruised, stale, and weary," resigned his presidency of the IDCA, a position he had held for five years, saying: "It now does appear that this form has become unsatisfactory to enough people that we should never try to stage a conference in this way again. While we have not learned from any individual or any of the dissenting groups what kind of conference they would like, it appears to me that it would be something so different from our past conferences and perhaps from our concerns with design that it must be put together with an entirely new vision if it is to continue."[93]

The other directors felt less depressed, and the majority voted that there should be another conference, after all. Richard Farson, Dean of the School of Design at the newly formed CalArts, shared his vision for what a radically redesigned conference might look like:

> I would like to run a high-risk design conference.
>
> Very dignified and sleazy, very specific and general. I would like to go both ways at once. I question the star system. I think we may need names to get them into the tent, but beyond that we don't need them. Reverse the flow of communication. ... It shouldn't be just informational. It should be mind-stretching. ... Should be more of a carnival.[94]

Farson—who, in addition to his CalArts deanship, was a psychologist and chairman of the Western Behavioral Sciences Institute, an organization involved in research on the leadership and communication of groups—believed it was important to redesign the conference from the bottom up. By introducing workshops, games, and other participatory formats, he wanted to bring IDCA in line with the new teaching methods being tested on campuses across America. After excitably enumerating his catalog of ideas, Farson concluded on a philosophical note: "I would like to say that any human grouping is vulnerable. We have a saying it is easy to damage an individual and not an institution. I disagree. We are very vulnerable as an institution."[95]

In fact, the institution of the International Design Conference in Aspen was battle-scarred but ultimately resilient. It remained under IDCA's leadership until 2005, when the American Institute of Graphic Arts assumed its administration. But the 1970 conflict did have consequences for the individuals involved. In a discussion between Noyes and Bass before the board meeting began (captured by Noyes's son's camera), Noyes appeared bemused and upset; he scratched his arms and his eyes wandered as he attempted to make sense of the palpable change in the conference atmosphere. "All those resolutions at the end had nothing to do with the subject of the conference," he said. "This is the politicizing—I believe that's the word—of the Aspen Design Conference. And I am not a political guy. I'm not interested in becoming a political guy. I'm interested in making my points through my work. I don't play games with this kind of thing. I just can't. It's not in me."[96]

It is possible that Banham, too, underwent a personal reevaluation during this conference as, right before his eyes, he saw the mood of the students turn against the Pop values he had championed in the previous decade. "So we didn't blow the conference," he commented, "but I count it among the hollower victories of my public career."[97] As we saw in chapter 1, hitherto Banham had been the hip spokesman for Pop, identifying emerging trends considered taboo by the design establishment. He had challenged the traditional design canon and introduced new subject matter for evaluation—from crisp packets to surfboards—and a new vocabulary for discussing them, which connected to the contemporary lexicon. Now that the qualities he had celebrated, such as expendability and styling, were being rejected by the younger generation of anticonsumerist designers, Banham found himself dislocated from the concerns of the counterculture. Furthermore,

2.17
Still from *IDCA '70* showing members of the IDCA voting against Eliot Noyes's motion to discontinue the conference at the IDCA board meeting, the morning after IDCA 1970. Dir. Eli Noyes and Claudia Weill, IDCA, 1970. Courtesy of Eli Noyes.

"CONFLICTING DEFINITIONS OF KEY TERMS"

his critical apparatus, which for the past decade, in the context of design magazines like *Design*, *Architectural Review*, and *Industrial Design*, and general interest publications like the *Listener* and the *New Statesman*, had seemed unconventional and even revolutionary, was, in this new critical environment of improvisational theater, petitions, and protest through resistance, no longer considered particularly relevant or effective.

The IDCA board had charged Banham with compiling a book. Even though *The Aspen Papers* was not published until 1974, tellingly, Banham's narrative ends with the 1970 conference which, he opined, "will be the last Aspen conference in anything like the form on which its reputation has been built."[98] He selected only two of the lectures from that year for publication, and put them under the title "Polarization." One was the French Group's statement, which he retitled "The Environmental Witch-Hunt"; the other was the urbanist Peter Hall's pragmatic presentation on English New Towns. Banham's preface to the 1970 section of the book, which draws attention to the gulf dividing "those who believed that rational action was possible within 'the system,' and those who wanted out," makes clear his preference for Hall's liberal reformist position.[99] In a draft manuscript for the book written in 1971, he wrote: "Aspen has survived (the right word, I hear) the conference of 1971 masterminded by the California connection (Jack Roberts, Richard Farson, Jivan Tabibian) and is girding its loins for 1972 under Richard Saul Wurman as Program chairman."[100] Because the book's publication was delayed, he rewrote the passage, but his ambivalence is still evident when he referred to the "extraordinary scenes of 1971—masterminded by Richard Farson under the entirely appropriate title of 'Paradox'"—and said that when he arrived for the 1973 conference, "Performance," chaired by Milton Glaser and Jivan Tabibian, "it would be an operation to which I was somewhat a stranger."[101]

While others were enthused by the way the conference was evolving in line with contemporary attitudes—and many saw 1970 as the beginning of something new—Banham's framing of *The Aspen Papers* is conspicuous for being a lament or a "memorial of sorts" for the conference and for a mode of criticism.[102]

Dissent at IDCA prior to 1970

IDCA conferences in the two decades prior to IDCA 1970 were not devoid of criticism, and the impetus toward consensus-building did not preclude dissenting voices, but since they were presented in the written form of pre-prepared lectures, IDCA was able to frame them on its own terms, and ultimately to absorb them. Among the forceful critics who had appeared at past IDCA conferences was the sociologist C. Wright Mills, who in 1958 had delivered a harsh critique of industrial design (discussed in chapter 1). In 1964 Dexter Masters, director of Consumers Union and editor of *Consumer Reports*, had spoken out about the conspicuous absence of serious criticism within the design industry, specifically with regard to the "corruption in designing that has the effect of economically cheating the buyer or endangering his health, or possibly his life, and insulting him as a fellow human being in the process."[103] In his lecture, "Quick and Cheesy, Cheap and Dirty," Dexter Masters described the work of his magazine *Consumer Reports* as exposing the various concealments practiced by both manufacturers and designers. He expressed his disappointment that designers did not want to write criticism in its pages, since it provided such ideal conditions: *Consumer Reports*, he told the audience, was objective because it employed scientific research (the magazine tested products in laboratories, and purchased them rather than being sent them by manufacturers and PR companies) and did not rely on advertising for revenue. Eliot Noyes had written design reviews for *Consumer Reports* between 1947 and 1960, but in recent years, when Masters asked Noyes and other industrial designers to review design products, they declined, refusing to comment on the work of other designers, seemingly under pressure from the Industrial Designers Society of America.[104]

Thus provoked by Masters, conferees at IDCA 1964 prepared a special document concerning the "failures of criticism," which outlined four resolutions intended to improve the situation. They called on design organizations, designers, manufacturers, and the media to take more responsibility for the encouragement of design criticism. Reyner Banham and Ralph Caplan (who had been editor of *Industrial Design* between 1958 and 1963) were both at the conference and helped to shepherd this initiative, which was subsequently published in *Industrial Design*.[105] Two of their resolutions addressed the unstated, but widely adhered-to, rules of the industrial design professional societies:

> Firstly—a lively interchange of well-informed critical opinion is essential to all branches of the business of design, and the professional bodies representing designers are strongly urged to encourage it.
>
> Secondly—designers have a duty to contribute their knowledge freely and honestly to public discussion of design in all its aspects. All restrictive rules which subject the public good to a narrow concept of loyalty to the profession by prohibiting designers from commenting on one another's work should be relaxed as soon and as far as possible.[106]

Furthermore, IDCA board members sometimes took aim at their own foibles and the shortcomings of the conference. In a 1969 speech, George Nelson painted a dystopian portrait of contemporary society as a staged play dominated by the military-industrial complex and tarnished by environmental pollution. In his vision, floating above the stage-earth were thousands of shiny, beautiful ice-blue spheres containing fusion material. Nelson noted how gratifying it was to see on these spheres a special credit for graphic design. "You may recall the decision made about a year and a half ago at the first international conference on orbiting garbage, when it was decreed that every aspect of man's environment should be studied by leading design professionals with the view to the ultimate beautification of everything," he said, clearly satirizing the limited and aesthetically focused objectives of the IDCA conferences which he helped to lead.[107]

These prior examples of criticism at the IDCA were articulated *within* the speeches of invited participants, and thus were contained in the accepted structure of the conference. Banham chose to publish two of these critiques in *The Aspen Papers*. In his preface to the section labeled "Dissent," he observed that "Aspen often called on the services of the more formidable social critics of the day."[108] By positioning these critics' commentaries as a commissioned "service," he was thus able to package them palatably alongside the more benign material. Even the 1964 attendees' resolutions, which arose spontaneously during the conference, were endorsed by the conference leadership and, because they took written form, were published in the conference materials.

2.18
Photograph of Ralph Caplan and Reyner Banham at IDCA 1964. Courtesy of Ralph Caplan.

What happened at the 1970 conference, by contrast, represented a more radical variant of criticism, one that was harder for the IDCA to assimilate. Houseman's complete lack of preparation for such a physical onslaught on the structures of the conference can be deduced from his opening remarks, in which he proposed the setting off of "some rhetorical fireworks, which I am sure are going to illuminate this congenial tent during the next five days."[109] He assumed that critique would be articulated only in the form of rhetoric as persuasive language, and would be contained in the prescribed arena of the conference tent. As we have seen from what actually transpired at IDCA 1970, written papers and moderated debate were only one option among many for conveying critique, which now included not just contributing to a conference, but also the more radical option of nonparticipation and resistance.

The 1971 conference as a response to the 1970 critiques

Just as he had outlined during the post-IDCA 1970 board meeting, Richard Farson used his chairmanship of IDCA 1971, titled "Paradox," to introduce more politically relevant themes and experimental communication formats. He picked up on the leftist thrust of the students' IDCA 1970 resolutions and the French Group's statement by attempting to address the major sociopolitical issues of the moment such as sexual politics, Third-World hunger, and what he termed the "revolution of consciousness"—an umbrella heading that allowed him to discuss the impact of drugs such as LSD. Sensitive to the 1970 conference attendees' critiques, he included a "Conference Feedback" session on the final day, during which the conference board members would "react to criticisms and comments about this year's conference."[110] There was only one main-stage speaker, the environmentally conscious architect-engineer R. Buckminster Fuller.[111] Other well-known figures, such as *Design for the Real World* author Victor Papanek, psychologist Milton Wexler, and *Born Female* author Caroline Bird, were presented as discussion leaders rather than keynote speakers, and there were fewer formal presentations and more roving, carnivalesque sessions, or "experiences," as Farson described them.[112]

On the first day of IDCA 1971, for example, Stanford University psychologist James Fadiman led a consciousness-expansion session in which participants explored their "transpersonal psychic states" through such techniques as "psychosynthesis." Later, Michael Aldrich, a member of the Critical Studies faculty at CalArts and coeditor of *Marijuana Review*, discussed "the role that drugs have played and will likely play in the history and future of civilization."[113] Other sessions saw an Esalen Institute staff member "enabling people to get in touch with the messages from their bodies," and "mythematician" Bob Walter conducting a game workshop in which participants explored "changing individual and social conceptions of sexuality, male-female balances, and the likely directions of sexuality in the seventies."[114]

Farson recalled that he wanted to introduce a new perspective on design to the Aspen attendees: "Design had been taught as a problem-oriented, Aristotelian thing. So designers would see things as problems rather than as predicaments. But in social design, we have predicaments, not problems. I wanted to show designers they were on the wrong track but also to be hopeful about the future."[115]

Other activities included video workshops led by the artist and CalArts faculty member Nam June Paik, meditation, a balloon ascent, design games, and the screening of films such as Kenneth Anger's *Invocation of My Demon Brother* and Thomas Reichman's *How Could I Not Be Among You?*. The artist and Associate Dean of Art at CalArts, Allan Kaprow, who wanted to counter "the whole concept of spectatorship," organized a "participation happening" using the Aspen ski lift and video technology.[116] Recalling the events of 1971, Farson said: "People had a chance to shape the situations they were in. They had a chance to affect the outcome and direction of the things they were participating in."[117]

The activities of IDCA 1971 can be understood in terms of the French Marxist philosopher Henri Lefebvre's theory of "moments."[118] A "moment" is an intense, euphoric, and ephemeral point of rupture in the flow of normal experience in which the possibilities of everyday life reveal themselves. Even though the moment passes and folds back into normality, the fact that something new has been exposed might have the capacity to change people's consciousness and help them escape their alienated condition. Through this critique of everyday life, Lefebvre hoped to refocus attention onto the body and the senses, and in so doing perhaps inspire the imagination to conceive of revolutionary or utopian possibilities.[119]

Utopie architect Antoine Stinco has observed how antimonumental, mobile, low-pressure inflatable structures, like those erected at IDCA 1970 and IDCA 1971, were a means of enacting Lefebvre's celebration of the festival of everyday life. Stinco explained: "The inflatable represented ... a festive symbol of the new energy. It did so through its fragility, its will to express the ideas of lightness, mobility, and obsolescence, through a joyous critique of gravity, boredom with the world, and of the contemporary form of urbanism that had been realized."[120] The moments of playful participation, a refocusing on the body, and the ethos of critique which characterized the fringe activities at IDCA 1970 conference and the core of IDCA 1971 conference can certainly be seen, in Lefebvrian terms, as ruptures in the smooth structure of the Aspen Design Conference and the principles of humanism, consensus, and pragmatism that underpinned it.

The list of books available in IDCA 1971's conference bookstore (also where the coffee was served, suggesting it would have received high traffic) covered a wide spectrum of contemporary thought. They ranged from feminist manifestos, such as Caroline Bird's *Born Female* (1969), Kate Millett's *Sexual Politics* (1969), and Betty Friedan's *The Feminine Mystique* (1963), to expositions on psychology such as Abraham H. Maslow's *Toward a Psychology of Being* (1962), Ken Kesey's *One Flew over the Cuckoo's Nest* (1962), and R. D. Laing's *Politics of Experience* (1967). The book list also encompassed Marxist texts

2.19
Photograph of TV monitor playing footage of attendees
on the Aspen ski lift as part of "Tag," Allan Kaprow's
happening, devised for IDCA 1971. Courtesy of The Getty
Research Institute.

"CONFLICTING DEFINITIONS OF KEY TERMS"

such as Herbert Marcuse's *One-Dimensional Man* (1964) and readings of Marxist influence on the rise of an American counterculture such as Theodore Roszak's *The Making of a Counter Culture* (1969). The list referenced recent thinking about communication such as Gene Youngblood's *Expanded Cinema* (1970) and Marshall McLuhan's *Understanding Media* (1964) and *The Medium Is the Massage* (1967), as well as politicized texts that inspired the civil rights and decolonization movements: *The Wretched of the Earth* (1961) by Frantz Fanon, *The Autobiography of Malcolm X* (1965), and George Jackson's *Soledad Brother* (1970).[121]

The list was largely based on the reading requirements of the Critical Studies department at CalArts, but it is also indicative of the kinds of literature that may have been read by the activists who had attended IDCA 1970. Theodore Roszak is helpful in highlighting the differences between the radical politics of students in Europe and those in America, which Baudrillard also identified, and gives some insight into the likely mindset of the students at Aspen. In *The Making of a Counter Culture* he characterized the young dissidents' worldview as a ragbag of philosophies. The countercultural young, he wrote, "are the matrix in which an alternative, but still excessively fragile future is taking shape. Granted that alternative comes dressed in a garish motley, its costume borrowed from many and exotic sources—from depth psychiatry, from the mellowed remnants of left-wing ideology, from the oriental religions, from Romantic Weltschmerz, from anarchist social theory, from Dada and American Indian lore, and, I suppose, the perennial wisdom."[122]

The "garish motley" to which Roszak referred was exemplified most literally in the garb of improvisational theater groups such as the Moving Company, present at both IDCA 1970 and 1971. Roszak's phrase refers figuratively to the eclecticism of the students' references, which ranged from the sociopolitically inflected design theory of Buckminster Fuller and Victor Papanek to Eastern philosophy.[123] Roszak's characterization is reinforced by Maurice Stein, Dean of Critical Studies at CalArts, who observed of the students: "They still read Fuller. ... They are reading Dubos, Goodman. They're reading Gary Snyder, they're reading the Whole Earth Catalog, they are reading the occult."[124]

Farson's emphasis on audience participation and on "new kinds of social architecture created to enable higher levels of interaction" extended to the planning of the conference. The registration materials included a matrix that outlined along the vertical axis "some old social institutions" such as "Marriage and Family" and "Learning and Schools," and along the horizontal axis some of the "social revolutions" such as "Communication" and "Sexual Politics." Registrants were invited to indicate the intersections that interested them the most.[125] Furthermore, Farson enlisted the help of the design students in his department at CalArts—"creating the conference was their project that semester."[126] He asked Sheila Levrant de Bretteville, a young teacher in the design department (who had recently returned from two years in Europe, where she had witnessed the student riots in Paris, and read Fanon and Marcuse), to create a publication onsite. Her solution was informed both by the need to be expeditious (she had a three-month-old son in tow) and

Cover of special edition of
the *Aspen Times*, designed by
Sheila Levrant de Bretteville
for IDCA 1971. Courtesy of
Sheila Levrant de Bretteville.

"CONFLICTING DEFINITIONS OF KEY TERMS"

2.21
Spread from special edition
of the *Aspen Times*,
designed by Sheila Levrant
de Bretteville for IDCA 1971,
including strips created by
attendees. Courtesy of
Sheila Levrant de Bretteville.

"CONFLICTING DEFINITIONS OF KEY TERMS"

by her interest in participatory and nonhierarchical publishing models. She handed out diagonal strips of paper to attendees and encouraged them to fill them with comments on the conference using handwritten and typed text, drawings, and Polaroids. On the last evening she collected the strips, pasted them up to form the pages of a newspaper, printed copies using the *Aspen Times* offset press, and delivered them to attendees the following morning.

De Bretteville's newspaper was inspired by the collective editorial process in publications such as *The Whole Earth Catalog*, which was published twice a year between 1969 and 1971, and assembled a plethora of tools, resources, and tips useful for a creative or self-sustainable lifestyle, haphazardly collaged using the visual language of a manual. She had also designed a special issue of *Arts in Society* about the genesis of CalArts in which she juxtaposed fragments of quotes gathered from the school with news photographs from the period, and put the contributors in alphabetical order and the contents page in the center, in order to create a wholly nonhierarchical and nonlinear publication.

Accordingly, in her newspaper for IDCA 1971, she chose not to prioritize the voices of the main speakers, but rather to "let the participants speak for themselves." [127] De Bretteville considered that her belief that graphic design could be "more than telling people when and how to get places" was "part of the zeitgeist," but it was also part of a distinctively feminist approach to graphic design that she was working out at the time. [128]

Farson's emphasis on the audience members' interaction with one another and the speakers became an embedded, and ratified, principle of future conferences. Still, Farson remembered thinking that his 1971 conference was "a mess." [129] Instead of having a closing speech, he had asked a guerrilla theater group to do a finale that would provide a summary of the conference. [130] "I guess they couldn't think of what to do, so what they did was to get miniature marshmallows and ran down the aisles and threw them at people. My heart sank, especially when afterwards I saw Elizabeth Paepcke [Walter Paepcke's widow, who continued to attend the conference each year] on her hands and knees peeling off these marshmallows from the floor." [131]

A conference framework, with its built-in need for purpose, preplanning, and a timetable, will always be an awkward social architecture for unprogrammed and genuinely participatory activity—and especially for critique directed against the host organization. The rupture at IDCA 1970 was truly spontaneous; the participants who stirred up the crowd believed in what they did and were excited at the possibility of change. The 1971 iteration of the conference, despite its vast array of group activities and its embrace of the social themes of the period, was more forced. No matter how creative Farson's ideas were for his "high-risk design conference," he was the ringmaster of the project, and in many respects had to follow IDCA protocol. [132] The consciousness-expanding sessions, which he favored, were fairly radical in terms of conference planning of the period, but the very fact that they were planned dissipated some of the energy of the corridor debates and heated resolution readings of the previous year.

"CONFLICTING DEFINITIONS OF KEY TERMS"

"Massive collusion under the sign of satisfaction"

The conference was promoted each year by a printed brochure, which was usually designed by that year's chair. In 1971, perhaps since Farson was not a designer, the IDCA board decided to use Noyes and Weill's film, *IDCA '70*, to announce the conference, even though it presented a message of critical dissent. IDCA president Jack Roberts (who took over when Eliot Noyes resigned) suggested that a two-minute epilogue or "commercial" for the 1971 conference be added to the end of the documentary. In a letter to Farson, he suggested that the epilogue "should say that's what happened, we're facing the climate-for-change and this is your invitation to come and participate in '71."[133] With this epilogue attached, ending with the briskly jovial line "We'll turn over a new leaf in Aspen!," prints of the film were loaned to design organizations and such corporations as Alcoa, Westinghouse Electric Corporation, and Whirlpool, which arranged screenings for their design departments. Seemingly there was much demand for the film, and IDCA quickly ran out of prints, advising some who wanted to see it to arrange group screenings.[134]

Instead of attempting to resolve the real issues that had been brought to light, the IDCA leadership subsumed them, demonstrating their similarity to the capitalist system's ability to assimilate its internal contradictions. As Baudrillard wrote of the events of May 1968, "When a system is able to stay in balance by blindly refusing to come to terms with a problem, when it is able to assimilate its own problems and even turn its own crises to advantage ... what is left other than to interrupt it by insisting on the almost blind need for a real pleasure principle, the radical demand for transgression, against the massive collusion under the sign of satisfaction?"[135]

2.22
Still from *IDCA '70* showing people inside an inflatable tent. Dir. Eli Noyes and Claudia Weill, IDCA, 1970. Courtesy of Eli Noyes.

The interruption of IDCA 1970's proceedings, seen as a manifestation of Baudrillard's directive, was indeed insistent in its "blind need" for the sensory pleasures of stumbling around in inflatable structures, play-acting, and picnics, and in its resounding "demand for transgression" of the prevailing institutional norms, but it was ultimately short-lived. The protest's lack of sustaining power was partly due to the simple fact that once the protestors left Aspen it was difficult for them to maintain the political energy generated during that week in June in the mountains, and to the wider reality of the declining energy of the counterculture throughout the 1970s.

CONCLUSION

Although in 1974, when *The Aspen Papers* was published, Banham felt despondent enough about the future of the conference to observe that "an epoch had ended," in fact, despite the intensity of the protests in 1970 and the experimentation with communicative formats in 1971, the IDCA did not implode, nor even irreversibly redirect its course. It absorbed the critiques leveled against it, and appropriated some of the new formats (it inserted a sanctioned space in the program for "conference feedback" and started calling speakers "resource people"), but eventually returned to a lecture-based structure with a bias toward celebrity designers, and continued to regard scholarship student attendees as the hired help while initiating a policy to cap the number of regular student attendees.[136]

As historian Matthew Holt has observed, Baudrillard regarded critique of a system as an essential function of that system's self-organization and triumph: "Critique in this sense is not a negation but an 'adjustment' to which the system responds ... (a thoroughly cybernetic vision of relations). Critique is now information in a system, not the presentation of a genuine alternative to that system."[137]

As the liberal design establishment folded the conflict of IDCA 1970 back into its thick blanket of consensus, the protestors and their various missions dispersed like cotton bolls in the wind. Some became the celebrity designers of future conferences, and many founded their own institutions, albeit alternative ones—Farallones Institute, the Environmental Workshop, Ecology Action, Esalen Institute, and CalArts, for example. In their new positions of responsibility, they faced their own internal contradictions as they attempted to balance a continued desire for transgression with the real needs of actual institutions. The critique directed at IDCA 1970 atomized, but the embodied methods and means of its enactment took root and, as this book will go on to explore, reemerged throughout the latter part of the twentieth century as the purposes, publics, and poetics of the written variant of design criticism became increasingly contested.

3.1
Photograph of Stephen Bayley, curator of The
Boilerhouse Project, 1981–1989, seated in
his office adjoining the gallery space. Courtesy of
Design Curial.

3

Designer Celebrities and "Monstrous, Brindled, Hybrid"
Consumers: The Polarizing Effects of Style
in the British Design Media, 1983–1989

One whole wall of the room was glass, allowing a viewer to see everything in it, including its occupant, the design curator. He was sitting on a black and chrome Mies van der Rohe chair, pulled up to a white desk illuminated by a black Artemide Tizio lamp, and accented by a white porcelain vase holding white flowers. He was sipping coffee, black, for "purity of vision," from a white Conran-endorsed Apilco cup edged with silver gray, and the handset of a British Telecom push-button telephone, resprayed to his specifications in dove gray, was expertly wedged between his ear and his Paul Smith-suited shoulder. As he talked, he turned the pages of *Blueprint*, selected from an angled shelf that ran the length of the wall behind him, displaying his "daily reading" of design, technology, and general interest magazines, among them *Forbes*, *New York*, the *Atlantic*, and *French Vogue*. Occasionally he made a note or two using his Mont Blanc pen; he might have typed them up later on his dark gray Olivetti ET121. Within hand's reach on the desk there was a spirit level, a toy Pininfarina Lancia Aurelia B20GT in original gray, four bottles of Tipp-Ex corrector, and a Falcon Safety Products Dust-Off canister of compressed air "for blasting dust away."[1]

   Stephen Bayley's office was such an exaggerated manifestation of cleanliness and order, such a carefully composed display of conspicuous consumption, that anyone would have been forgiven for thinking it was one of the prescriptive exemplars of "good design" on show in the Boilerhouse Project gallery it adjoined. As we shall see, in mid-1980s design commentary and curation, it was getting increasingly hard to tell where the line lay that may once have separated inside from outside, reality from image, and treasure from trash.

INTRODUCTION

Style wars

In early 1980s Britain, and specifically London, the design media began to celebrate the nebulous but pervasive concept of *style*, in the sense of the considered expression of personal identity and social position through consumer goods and lifestyle choices. The topic of style held a particular and personal fascination for some of the editors, writers, and curators who orbited the independent design publication *Blueprint*, launched in 1983, and the privately funded design exhibition center The Boilerhouse Project, opened in 1981. Among those who chronicled this phenomenon were the journalist Deyan Sudjic, management consultant Peter York, and curator Stephen Bayley. Their engagement with this subject matter aligned with a definition that considers style "not an *aspect* of things, people, or activity," but rather, what "constitutes them as what they are."[2]

Blueprint and The Boilerhouse, and the writers and curators associated with them, can be seen, in Bruno Latour's phrase, as having "gathered" new publics for design commentary and curation.[3] In their roles as public sites of debate and exchange, the magazine and the exhibition center opened up discussion of design to an expanded audience and helped to map, label, and define the unfamiliar territories of design, style, and taste during a period of intense, but often confused, interest in such topics. While the kinds of commentary about design and style found in the pages of *Blueprint* and in the Boilerhouse gallery was sophisticated, engaging, polemical, but largely apolitical and nonideological, some design criticism of the period offered a darker interpretation of style and the lifestyle "crazes," diagnosing them as symptoms of a social pathology in need of remedy.

Left-leaning critics, who were interested in design and consumer culture as subject matter, pointed to the more troubling aspects of the way in which the media conflated design and style with status during the Thatcher era. They identified design practice and design commentary as perpetuating the values that led to the irresponsible use of credit, the concealment of class and racial schisms, society's separation from the means of production, and its abandonment of communal values. Among such critics were the socialist cultural critic Judith Williamson and the cultural theorist Dick Hebdige, who wrote about design, style, and consumer culture in Leftist publications such as *Marxism Today* and *New Socialist*, academic journals such as the visual culture journal *Block* and the photography theory journal *Ten.8*, as well as in more widely read magazines such as the social issues magazine *New Society*, the style, music, and fashion magazine *The Face*, and London listings magazines *Time Out* and *City Limits*. Williamson, among others, warned against the way a design and style rhetoric was being wielded both by the Right to conceal social and economic problems such as unemployment, unrest in inner cities, anti-trade union legislation, and continuing violence in Northern Ireland, and by the Left as a seductive distraction from internal rifts in its organizing parties and inability to provide a powerful alternative to Thatcherism.

Using a selection of their articles that dealt explicitly with style and lifestyle, this chapter looks at how Hebdige and Williamson situated design in a multidimensional context, incorporating discussion of the emotional needs of the consumer and the role of class, gender, and identity in ways that, although marginal at the time, anticipated the direction in which some design discourse was to develop in the 1990s and beyond.

Design as an economic weapon and a "key to national salvation"[4]

Design played such a visible role in the credit and consumer boom of 1980s Britain—through interiors for the fashion retail revolution in the high street, identities for the new corporate mergers and privatized public utilities, and in design-referencing television commercials—that commentators referred to the 1980s as the "design decade," even before it was over.[5] Design consultancies like Michael Peters and Fitch & Co. went public, diversified their services, and grew their ranks to many hundreds.

The Conservative Party, with Margaret Thatcher as prime minister, which came into power in 1979 and was reelected in 1983 and 1987, supported design's entrepreneurial growth, seeing it as a key weapon in the economic strategy for a more prosperous Britain.[6] In 1984, for example, the government granted £20 million to the funded consultancy scheme, which paid the design fees for a small or medium-sized manufacturer who hired a design firm, and in 1986 further demonstrated its commitment by spending £12 million on design initiatives.[7] A 1984 *Blueprint* editorial enthused:

> It has been a remarkable year for the world of design and architecture. In Britain, the design boom has accelerated, with clients, being prepared for the first time, to take industrial design seriously. The design consultants are chasing each other onto the Stock Exchange in pursuit, one suspects, of prestige as much as investment. Mrs. Thatcher's government has increased its backing for design considerably. It has doubled the sum allocated for the funded consultancy scheme to £20m, in effect providing any manufacturer with a free sample of the potential of a design consultant.[8]

Thatcher held lunches for designers at Downing Street and appointed John Butcher as the first Minister for Design, with a mandate to engage design in service of the national interest. In a 1987 speech, Butcher praised a selection of design companies: "Here is design at work. Improving competitiveness. Winning markets. Increasing profitability. ... That's what design is about."[9]

The most visible expression of the era's credit boom was to be found in retailing. High-street chain stores, banks, and boutiques, such as Next, Richard Shops, Midland Bank, Esprit, and Joseph, used design to give them a competitive edge and to create dramatic settings for consumption. They employed interior design firms, such as Fitch, David Davies Associates, and Eva Jiricna to design their interiors, "in styles ranging from candy-box Memphis to high-tech," and graphic design firms like Why Not Associates to design their New Wave catalogs, signage, and logos.[10] "A centrepiece of this new retailing is design," wrote Robin Murray, chief economist of the Labour Greater London Council, in *Marxism Today*: "Designers produce the innovations. They shape the lifestyles. They design the shops, which are described as 'stages' for the act of shopping. There are now 29,000 people working in design consultancies in the UK, which have sales of £1,600m per annum. They are the engineers of designer capitalism. With market researchers they have steered the high street from being retailers of goods to retailers of style."[11]

Robin Kinross, graphic designer and author, identified graphic design—its ability to rapidly generate new identities, and fashionable skins for stores and corporations alike—as the defining design practice of the era. In a 1988 overview of postwar graphic design written for *Blueprint*, Kinross announced the simultaneous aggrandizement and implosion of the discipline in 1980s Britain:

> The most remarkable development in graphic design in the 1980s has been the process by which everything aspires to the condition of graphics: not just print or screens, but architecture, interiors and products. As working parts dematerialize into slivers, so the false fronts, pastel shades and "Matisse effects" take over. Or, seen another way, the developed world becomes a weather-free shopping arcade, with beggars kept on the move. This might as well be the apotheosis of graphic design, as it explodes into "marketing design," "retail design," or whatever term settles. Design becomes inflated into a way of life, a key to national salvation.[12]

Design and style in the public eye

Several new publications dedicated to design, advertising, and marketing, such as Marketing Week Publications' *Creative Review*, launched in 1980, and *Design Week*, launched in 1986, joined existing design-focused magazines on the newsstand such as *Design*, *The Designer*, and the *Architectural Review*.

New television programs also helped cement design and designing as subjects of general interest beyond the concerns of the profession. In 1981 the BBC launched a series of Horizon programs called "Little Boxes," about design and scientific thinking, written and presented by Stephen Bayley and directed and produced by Patrick Uden (with research by Penny Sparke) and featuring interviews with designers such as Raymond Loewy, Dieter Rams, and Ettore Sottsass. In 1986 the "BBC Design Awards," presented by the media personality Janet Street Porter, attempted to engage viewers by inviting people to vote for their favorite example of contemporary design. The BBC also began a "Design Classics" series in 1987, produced by Christopher Martin and commissioned by Alan Yentob, with thirty-minute episodes devoted to the VW Beetle, Sony Walkman, Barcelona Chair, Coca-Cola bottle, and Levi's 501 jeans. Meanwhile other channels developed their own design coverage. In 1984 London Weekend Television's "Hey Good Looking" series, produced and directed by Kim Evans and Bob Lee, and screened on Channel 4, devoted twenty fifteen-minute programs to four subjects—Style, Architecture, Design, and Advertising—written and presented by Peter York, Deyan Sudjic, Stephen Bayley, and Janet Street Porter, respectively. In their focus on what were beginning to become the hackneyed icons of design and celebrity designers, such programs enforced the conventions of design commentary. "Design Matters" was a more ideas-oriented ten-part series launched in 1984, advised by Ken Baynes, and commissioned by Channel 4 from the production company Malachite. According to media theorist Paul Springer, the show saw high audience figures, in the 500,000s, due to its focus on design as it affects everyday life. A typical episode juxtaposed a fruit and vegetable storeowner organizing his display, alongside Sainsbury's supermarket organizing theirs, helping to expose design and its decision-making and planning processes as part of everyday life.[13]

In addition to television's instrumental role in increasing awareness of design, many newly launched youth culture and style magazines such as *The Face*, *Sky*, *Blitz*, and the men's magazine *Arena* also covered design alongside their staple fare of music and fashion. Style, in these magazines, was mainly understood in terms of the staging of fashion shoots, which included casting, selecting clothes and accessories, hairstyling, and the directing of facial expression and body stance. This activity was most famously encapsulated in the work of stylist Ray Petrie, who introduced the Buffalo style with sultry-looking teenagers, often of mixed racial descent, wearing designer clothes that he paired with underwear, vintage pieces, and athletic basics. In these magazines, style was also synonymous with youth. In her analysis of the emergence of "youth" as a distinctive market segment during the 1980s, sociologist Celia Lury refers to the way "consumer culture provides an environment in which age—specifically what it is to be young—is constituted as a style rather than a biological or even generational category."[14] Style, as depicted in youth culture magazines, was free from any ideological moorings, and thus available for appropriation in other media contexts such as *Blueprint* magazine or The Boilerhouse Project.

These youth culture magazines also covered design through the lens of style. The May 1985 issue of *The Face* featured pieces on contemporary architecture, Nigel Coates, Ron Arad, and the magazine's own art director Neville Brody, while the June 1985 issue featured a profile of Terence Conran, and the end-of-year round-up issue featured Philippe Starck, for example. Reviewing the year 1984, Robert Elms characterized style "as our national sport" and averred that "vases, staplers, and coffee makers" were the "real signs of the times."[15]

"Style's become the new language"[16]

Peter York, style editor of *Harpers & Queen*, management consultant and marketing specialist, began his episode of the television program "Hey Good Looking" with the pronouncement: "Style is the way things look and the way they are."[17] According to York, who, with a pop sociologist's eye, tracked the struggles between the various youth culture tribes, style in early-1980s Britain was both "the most difficult word in the language" and "the new language."[18] Style could be used to "express not just who you are, but who you'd like to be," and was accessible thanks to a profusion of images spread by magazines, film, and television.[19]

This characterization of style as a consumable and interchangeable element of personal identity jarred with the traditional art-historical understanding of the term at the time. In art history, style is generally considered to be an analytical tool and a descriptor, a means of identifying and labeling the work of an artist or an epoch on the basis of its visual characteristics. The art historian Meyer Schapiro, known for his mid-twentieth-century study of style as a diagnostic tool, propounded the formalist definition: "By style is usually meant the constant form."[20] His belief in the "constancy," rather than the variability, of style represents the prevailing view of the term in most art discourse for much of the twentieth century.[21]

In the 1980s, transposed to design discourse, the term style still bore residues of these definitions—some design journalists and critics, including Stephen Bayley, were trained as art historians, after all—but the term had accrued new meanings, partly through its popularization in the media. Design discourse was also still negotiating its modernist ancestry where style meant the surface appearance of a product—a quality that was held in implicit contrast to its substance or function. As we have seen, the term style had accrued derogatory connotations during the 1950s and 1960s, through its association with the planned obsolescence techniques of the Detroit automobile industry. Richard Hamilton's and Reyner Banham's fascination with the glamorous qualities of American image art direction and the symbolism of popular products, discussed in chapter 1, anticipated to some extent the enthusiastic embrace of style in the 1980s. Banham's and Hamilton's analyses were focused more on the style of objects and images as physical entities, however, while by the 1980s design commentary had begun to be attuned to more ephemeral and body-centric manifestations of style such as clothing and accessories.

Another shift in the meaning of style can be deduced from the fact that, while historically style was a measure of a culture, in 1980s Britain it was increasingly associated with the individual and individualism, a narcissistic condition that had been wryly assessed by American journalist Tom Wolfe, in his article "The 'Me' Decade and the Third Great Awakening," as a millenarian outburst of vitality, and castigated by cultural historian Christopher Lasch in *The Culture of Narcissism: American Life in an Age of Diminishing Expectations* as a social pathology and a decadent defiance of nature and kinship.[22]

Complicating this etymological evolution still further, the term *lifestyle* began to be used interchangeably with, and even to supplant, the term style in design discourse of the 1980s. Lifestyle reflected a set of values that extended to leisure activities such as where one dined, what one read, and how one appointed one's home. It had reached the high street in the physical form of House of Fraser's "Lifestyle" shop, a boutique section of the department store, selling tableware, fashion, menswear, furniture, and lighting, aimed at twenty-five-to-forty-five-year-olds and newlyweds. In its review of the year 1985, *The Face* magazine reported: "the designer lifestyle became the mass-market lifestyle via Miami Vice. The aesthetics of consumer goods became a subject of intense interest as architects became window dressers and artists became interior designers. If you couldn't change the world this year, at least you could change your curtains."[23]

What lettuce can say about you

In 1979 the French sociologist Pierre Bourdieu published an analysis of the relationships between "the universe of economic and social conditions and the universe of life-styles" in contemporary French society.[24] *Distinction: A Social Critique of the Judgment of Taste* was translated into English in 1984, and was quickly adopted by many British cultural theorists as a landmark text and a helpful model in their own efforts to understand the ways in which taste was being used as a tool for establishing and maintaining class distinction. It also helped to focus intellectual attention on the notion of lifestyle as subject matter, and as a site of mutual exchange between producers and consumers. Bourdieu used the term "habitus" to mean "the represented social world," or "the space of life-styles" as it was constituted by the relationship between "the capacity to produce classifiable practices and works, and the capacity to differentiate and appreciate these practices and products (taste)."[25]

Celia Lury identifies the emergence of "lifestyle as the definitive mode of consumption" in the 1980s.[26] She draws both on Bourdieu's observations of the socially patterned nature of taste and on Dick Hebdige's characterizations of the ways in which subcultures used style choices to define themselves, and how objects and surfaces and the "signifying practices which represent those objects ... render them meaningful."[27] Lury defines lifestyle in the 1980s as a "new consumer sensibility" through which people sought to symbolically and aesthetically display their individuality and their sense of style through their choice of a particular range of goods.[28]

The notion of a designer lifestyle being available for purchase had emerged in the 1960s through the design-led retail enterprises of Terence Conran in Britain and Ben Thompson's Design Research store in Boston in the US. As Jane Thompson observed, the store "pioneered a new way of buying for life—an integrated way of thinking about your life."[29] Thompson and Conran were designers and restaurateurs as well as retail entrepreneurs, thus reinforcing Bourdieu's portraits of the parallels between people's literal taste—what they like to eat—and their aesthetic taste in household furnishings:

> Before lifestyle was a buzzword, architect Ben Thompson (1918–2002) sold it—his version of the most up-to-date way to live—from a clapboard house on Brattle Street in Harvard Square. His store, founded in 1953, was called Design Research, and what it sold was a warm, eclectic, colorful, and international version of modernism, one that mixed folk art and Mies van der Rohe, Noguchi and no-name Bolivian sweaters, offering newlyweds and Nobel Prize winners one-stop shopping for tools to eat, sleep, dress, even to party in a beautiful way.[30]

By the 1980s, however, lifestyle was operating at a more industrialized scale, and was increasingly pivotal to marketing strategy and product development in companies such as Sony, which launched a range of niche-market Walkman products starting in 1979. The lifestyle phenomenon had also gained visibility due to the increasing numbers of publications, commercials, and television programs that featured the "dispositions," lifestyle choices, working environments, processes, and habits of designers.

In all varieties of discourse of the period—from youth culture magazines to Leftist mouthpieces—lifestyle was portrayed as a means of signifying one's identity, the attributes of which could be studied and learned, and the products of which, for those who could afford them, were readily available for purchase. In its most positive conception, consumption of the designer lifestyle offered a way to cut across class boundaries at a time when they were already shifting due to the rise of an expanded middle class in the flux of a post-Fordist economy.[31]

The Face characterized the phenomenon more cynically as a top-down imposition: "From Adlands ideal home (Montblanc pen, Braun calculator, Tizio lamp) to Sainsbury's fresh food counter (why buy Radicchio rather than Iceberg? Kos it says much more about you than ordinary lettuce can). ... Design is everything and everything is design."[32] Meanwhile, *Marxism Today* presented the lifestyle craze as marketing's efforts to catch up with working-class behaviors: "The buzzword is lifestyles—a concept which goes hand-in-hand with the retail revolution. Lifestyle advertising is all about designer-led retailing which reflects changing consumer demand. In essence it is marketing's bid to get to grips with today's social agenda—the changing shape of working-class culture, the impact of feminism, ethnic spending power, the 'new man'—all these identities are up for grabs in lifestyle campaigns."[33]

And yet, as critics like Judith Williamson argued, merely buying the accouterments of a designer lifestyle didn't bring you any closer to being a designer. In her view, ideologies such as consumer fads were still firmly tied to, rather than cut loose from, the economic realities of people's lives: "The possession of expensive jogging shoes, videos, home computers and so on does not necessarily mark a level of fulfilment for the supposedly right-wing 'bourgeoisified' working class but, in part at least, a measure of frustration. Their aspirations have been caught up in the wheel of consumer production. Wearing a Lacoste sweatshirt doesn't make anyone middle class any more than wearing legwarmers makes you a feminist."[34]

Other commentators also suspected that the sense of choice brought about by an expanded consumer culture—York's belief that "the whole world, past and present, is your dressing-up box"—was actually limited to a prescribed set of options, and ultimately illusory. As Social Democratic Party member David Marquand observed of social behavior during Thatcher's term, "The range of identities legitimized by the enterprise culture is very limited. It gives increased scope for one's identity as a consumer, but not to other identities. Indeed, it is positively hostile to identity-choices that threaten the authority of the entrepreneur and the supremacy of entrepreneurial values."[35]

Similarly, Williamson believed that the enterprise culture used consumerism to distract the public from ever-widening social and economic inequalities, writing in 1985: "it is precisely the illusion of autonomy which makes consumerism such an effective diversion from the lack of other kinds of power in people's lives."[36] Such critiques will be explored in more detail in the latter part of this chapter.

Illness as metaphor

Some British critics in the 1980s cast the viral attributes of style as an infecting property, and the cause of sickness. Clinical metaphors abound in their writings of the period. In her 1978 essay "Illness as Metaphor," the American novelist and essayist Susan Sontag discussed the way in which the diseases tuberculosis and cancer were "encumbered by the trappings of metaphor," closely related to the economic practices of the periods to which they are connected. She characterized the nineteenth-century disease TB as being linked to "consumption" and "wasting," and the twentieth-century disease cancer as being linked to "abnormal growth" and "refusal to consume or spend."[37]

Critics of excessive consumption in the 1980s, such as Williamson and Hebdige, shared Sontag's negative view of capitalism's dependence upon "the irrational indulgence of desire," but continued to use the metaphorical imagery of sickness, nevertheless.[38] It was as if the symptoms of Sontag's characterization of the nineteenth-century condition of TB—the wasting, the inability to gain nourishment from consumption—had reemerged even more vigorously in late-capitalist society but were now also applied to a metaphorical conception of society's moral character and mental state. In a second essay, written in 1989, Sontag turned her attention to the stigmatizing effects of military metaphors used to describe AIDS, a disease which had come to public attention in the mid-1980s. She looked at the way in which individual cases were transposed onto society as a whole: "AIDS seems to foster ominous fantasies about a disease that is a marker of both individual and social vulnerabilities. The virus invades the body; the disease (... or the fear of the disease) is described as invading the whole society."[39] Similarly, in design commentary of the 1980s, the style sickness was often identified as not merely an individual affliction, to be attended to on a local level, but rather a national epidemic necessitating sweeping social and political reform. Martin Pawley, an architecture critic, reflecting in the *Weekend Guardian* on the relentless assault of home furnishing companies, wrote: "[Home] isn't safe, it is infested with the virus of consumption. The vast flood of credit it generates enables it to play the part of the Swiss sanatorium in the drama of the last stages of your disease."[40]

Compulsive shopping, shopping addiction, shopaholism, and compulsive buying (CB) are all terms that began to be used during this period. 1994 saw the start of a two-year British study of shopping addiction, funded by a grant from the Economic and Social Research Council, "to establish if there is a 'continuum' of shopping, running from normal purchasing, through impulse buying or binge shopping into full-scale addictive behaviour."[41]

A December 1988 piece in the *Guardian* talked of "symptoms," "design's current state of *hysterical* self-indulgence," and "the present design *mania*," alluding to ideas in Freudian psychoanalysis.[42] Its authors, the writer Jon Wozencroft and graphic designer Neville Brody, extended this set of references into the biological realm when they posited that "Style is a Virus" and that, "like design, 'style' is now a badly infected word."[43]

Hebdige, who noted a growing interest on the part of psychotherapists in treating the "sick at hearth," also forcibly forged the connection between psychosis and the contemporary preoccupation with style, and even pathologized his own writerly voice.[44] Many of the symptoms of the psychotic state Hebdige associated with consumerism found their equivalents in the physical form of his essay, which will be discussed in detail later in this chapter.

The title of Williamson's book *Consuming Passions*, and the themes of her collected articles therein, suggested that the greater our focus on individualism and away from communal values, the greater our appetite for consumer goods, the more we waste away, like victims of diseases like consumption or bulimia nervosa.[45] In an article titled "Anorexia of the Soul," York depicted the baby-look phenomenon, where adults—once hippies, now yuppies—dress like children to avoid having to confront guilty feelings about embracing capitalism. He painted a harsh portrait of "Little Mo," a female of the species regressed into infantilism, with "a big brown Just William satchel with a sticker of Rupert Bear on it."[46]

The pathological metaphor was also extended to the activity of criticism itself. Hebdige used a surgical metaphor to describe his perception of the shifting role of the critic. Speaking of the philosopher Jean Baudrillard's use of negation as a tactic, he wrote:

> Psychosis, waste and death are positively valued so that only "fatal strategies" can prevail. A "negative" cultural tendency is countered not in "resistance" or "struggle" (the terms of dialectic) but in a doubling of the same: a "hyperconformity" or hypercompliance. The critic-as-surgeon cutting out and analyzing diseased or damaged tissue is replaced by the critic-as-homeopath "shadowing" and paralleling the signs of sickness by prescribing natural poisons which produce in the patient's body a simulation of the original symptoms.[47]

In their engagement with style, the magazine *Blueprint* and the exhibition space The Boilerhouse can be seen, in Baudrillardian terms, as spaces of "homeopathic" commentary rather than of "surgical" criticism. They performed less as precision instruments calculated to "gouge out the rot," and more as simulating mirrors of a condition, "shadowing and paralleling the signs of sickness" in the design community and beyond.[48] These ideas will be further explored later in this chapter. At this point it is necessary to provide some background on *Blueprint* and The Boilerhouse Project as primary protagonists in shaping style-centric design discourse in 1980s Britain.

PART ONE: *BLUEPRINT* MAGAZINE'S USE OF STYLE TO RESET THE STAGE FOR DESIGN DISCUSSION

London's magazine of design, architecture and style

Blueprint, a large-format and image-rich publication about design and architecture, was launched in October 1983 with an extravagant party in the almost-completed Lloyd's Building. The magazine was published by the architect and editor Peter Murray (through his company Wordsearch Ltd), edited by the architecture and design journalist Deyan Sudjic, and designed by Simon Esterson, art director of *Designer's Journal*, published by the Architectural Press.

Sudjic believed there was a market share gap for a magazine like *Blueprint*, which would address both architecture and design, and would distinguish itself through the addition of style as subject matter. For the first year of its existence, *Blueprint*'s all-caps titular subhead was "London's magazine of design, architecture and style." "Style" gave *Blueprint* a thematic device with which to cut across typically segregated disciplines and open up design to a broader public by connecting it to a popular concern of the period. *Blueprint*'s conception of style evoked the calculated activities, products, and attitudes of an elite subset of London's design culture, in relation to which *Blueprint* positioned itself as both an insider and an interpreter.

3.2
Deyan Sudjic in 1979 receiving the award for young journalist of the year by the International Building Press. Courtesy of International Building Press.

3.3
Second issue of *Blueprint*, November 1983, featuring Daniel Weill, the product designer, photographed by Phil Sayer. Courtesy of Design Curial.

3.4
Photograph of 26 Cramer Street, home of the
Blueprint office, as well as the 9-H Gallery.
Courtesy of Peter Murray.

3.5
Photograph of 26 Cramer Street, home of the
Blueprint office, as well as the 9-H Gallery.
Courtesy of Peter Murray.

Sudjic recalled: "we were pushing back against the *Architectural Review*, which was spectacularly dull. It featured dull buildings and the writing was didactic, careful, polite, good taste—dull."[49] Sudjic also saw a void to be filled in product design coverage, since *Design* magazine, where he had himself worked as assistant editor under the editor-in-chief John E. Blake in the late 1970s, was, by 1983, "so tedious. ... It had lost faith in itself and had no direction." The Design Council, according to Sudjic, "came from a strand of thinking that design was something the well-bred inflicted on those that had no choice, which seemed deathly and, once the well-bred lost any sense of what they believed in, then they were a fantastic Aunt Sally to target and be very rude about. It was a gift."[50]

Indeed, *Design* in this period did follow a somewhat formulaic pattern, with features categorized baldly according to their industrial types—tableware, ceramics, appliances, and furniture, for example. The thematic articles, with titles such as "Should Products be Decorated?," seemed still to be grappling with the preoccupations of an earlier era.[51] The May 1983 issue, for example, included an article on Scandinavian furnishings, a special survey on plastics, and an analysis of the Duracell flip-top Durabeam torch, which was introduced with an unquestioning acceptance of its success, "financially, functionally, and aesthetically."[52] The issue's coverage of fashion was confined to a report from the very dismal-sounding "International Slipper and Footwear Fair in Blackpool."[53]

Sudjic, Murray, and Esterson assembled *Blueprint* after-hours from their day jobs, capitalizing on their experience and contacts at other publications but keen to create a distinctly new magazine over which they could exert complete aesthetic and editorial control. They drew on influences from a wide range of publications, including: Le Corbusier's avant-garde early-twentieth-century *L'Esprit Nouveau*; the Italian architecture magazine *Domus*, with its designer-as-editor philosophy; the celebrity-focused *Harper's Bazaar*; the caustic *Private Eye*; and, later in the decade, New York's satirical celebrity magazine *Spy*.

To begin with, the magazine's headquarters was in the Architectural Association's Communications Unit in Bedford Square and, after 1984, in the Putney offices of the graphic design firm Minale Tattersfield, where once a month, on Saturdays, Murray, Sudjic, sub-editor Jane Hutchings, Esterson, and a couple of graphic designers congregated to put the magazine together. Later they moved into a building at 26 Cramer Street, in Marylebone, which was awaiting development and could be rented for £4.00 a square foot (about one-fifth of market rate in that area). *Blueprint* shared the building with the 9H Gallery, the architectural firm Armstrong and Chipperfield, and the designer Sebastian Conran. Murray recalled: "In spite of the unprepossessing look—*The Daily Mirror* called it a 'rat-infested slum'—it was a key location in the 80s cultural scene."[54]

The magazine's launch was funded by a group of leading architects and design consultants including Terence Conran, Terry Farrell, Rodney Fitch, Norman Foster, Marcello Minale, Michael Peters, and Richard Rogers, who each contributed between £1,000 and £2,000 and would go on to feature as advertisers, contributors, and the subjects of articles. The magazine was sold for 95p on newsstands and via subscriptions, but its main source of funding came from full-page advertisements sold to contract furniture and lighting companies, furniture showrooms, and construction materials fabricators, such as Casa and Design, The Lighting Workshop, Bristol International, Artemide, Intercraft office systems. Murray and Sudjic raised more money by taking advantage of the Enterprise Allowance Scheme, a Thatcherite initiative in which investors in a business would be guaranteed a tax write-off if their prospect failed. The magazine prospered, and by 1989 Sudjic was forced to acknowledge: "I've become an entrepreneur in the Thatcher revolution ... a company director."[55]

Blueprint's writers and readers

The writing published in *Blueprint* was more ambitious and self-consciously literary than most design journalism at the time. The magazine prioritized its writers' sensitivity to language, included reviews of books like Nicholson Baker's *The Mezzanine*, and used literary and cultural references to add dimension to news pieces. In the approach to the year 1984, for example, references to George Orwell's novel *1984* were plentiful and, in a piece touting an upcoming "Designers' Saturday" event, Sudjic wondered if furniture purveyor Walter Collins knew "that one of his intimates, John Le Carré, refers to a 'posh light' (probably the famed Tizio) in his paean to the Israeli secret service, *The Little Drummer Girl*?"[56]

In his role as editor, Sudjic was keen to bring to the magazine's pages as many divergent voices as possible, to represent both sides of the modernist–traditionalist debate which dominated architectural discourse of the time. Between 1983 and 1989 the most frequent contributors were Janet Abrams, Colin Amery, Gillian Darley, Jonathan Glancey, James Woudhuysen, Martin Pawley, and, in the later years of the decade, Rick Poynor, who became deputy editor in 1988. *Blueprint* offered these writers a space for serious and skeptical reflection on design and its meanings. As Robin Kinross has recalled of contributing to *Blueprint* in the late 1980s: "I felt I was part of some sort of intellectual scene, in conversation with writers (Jan Abrams, Brian Hatton, Rowan Moore come to mind first) who were way beyond the hack journalists in powers of thought and expression, and in their intellectual reach."[57]

By the late 1980s *Blueprint* estimated that its paid circulation was 7,500 and its "pass-on readership" around 30,000. Two-thirds of its readers were architects and interior designers; the remainder were "other designers."[58] The first three issues were distributed only in London, but the magazine increased its distribution, and in 1986 it heralded itself as "Europe's Leading Magazine of Architecture and Design."

In her study of the early-twentieth-century American women's magazine *Ladies' Home Journal*, the historian Jennifer Scanlon identifies a contradiction between the educated and professionally satisfied women who edited and wrote articles in the *Journal* and the disenfranchised middle-class housewives who read them.[59] Junior designers and architects in 1980s Britain may have felt a similar disconnect between the everyday reality of their mundane jobs and the glamorous parties they read about in *Blueprint*, which created the impression of behind-the-scenes access to the stars of the design world. Even *Blueprint*'s initial editorial meeting took place in a private room at the design-conscious and celebrity-favored restaurant L'Escargot in Soho. The magazine referenced the habits and proclivities of designers with an easy familiarity, and used titles for its articles that exaggerated the social nature of interview appointments. Janet Abrams's article on Andrée Putman was titled "My Tea with Andrée." Headlines that emphasized privileged access to designers, such as "Maurice Cooper meets ..." or "Deyan Sudjic talks to ..." helped to draw readers into seemingly exclusive conversations which they, vicariously, could feel they were a part of.

Simon Esterson, *Blueprint*'s art director, was both a producer and consumer of design and a typical reader of the magazine. As Peter York put it, "Designers live in and through magazines, on the colour-printed page. History is fifties collectors' issues of *Look* and *Harper's Bazaar*. The play of Ideas comes through *Zoom* and *The Face*. ... Designers are magazine freaks and a half."[60] Esterson collected back issues of *Architectural Review* and found in the art direction of *The Face*, *Arena*, and *Skyline* inspiration for his own work at *Blueprint*.[61] In Sudjic's opinion, Esterson was "the intellectual conscience of the [*Blueprint*] operation. He had a huge range of references, from Constructivism to the history of the Architectural Press, and made me think about what we were doing in a way you don't as an innocent."[62]

A figure like Stephen Bayley might be considered another of *Blueprint*'s typical readers. He was the subject of several reviews, profiles, and gossip items in its pages. He also contributed articles, most controversially his takedowns of such antimodernist national treasures as John Betjeman and William Morris. And he clearly read the magazine, or at least parts of it, since his letters, protesting perceived inaccuracies or unfair critiques of his work, were published so regularly they provoked Fiona MacCarthy to comment: "I hope it is not a reflection of the paucity of material available to your correspondence columns that every issue of *Blueprint* seems to include a letter from Stephen Bayley."[63] In a profile on Charles Jencks, written by Sudjic, Jencks was quoted saying of Bayley: "that's what's so unspeakable about that taste show by Stephen Bailiff. It's low kitsch, he's trying to become an arbiter of taste ... and that's repressive, vulgar and in terribly poor taste."[64] Bayley wrote in to retort: "The first trouble with Dr. Jencks, to say nothing of the hypocrisy when it comes to castigating self-styled ambitious tastemakers, is that he and his supporting circus of superannuated hippies think that architecture is *only* style, rather as if it were pop music or coiffure."[65]

This kind of verbal sparring between major players in London's design and architecture world, which tracked across the issues of *Blueprint*, must have made for compelling reading. One can imagine readers receiving the latest issue with anticipation of its lively content, and turning first to the letters pages to see what new controversy might be afoot.

Blueprint's ontology of style

Style manifested in the pages of *Blueprint* in several guises. Most explicitly, it meant fashion and fashionable living as subject matter: articles about clothes retailing; profiles of designers such as Rei Kawakubo, Katharine Hamnett, Paul Smith, and Issey Miyake; and articles on nightclubs, restaurants, and boutique hotels. Secondly, it was both an overt aim—the founding editorial said the magazine intended to "keep a sharp eye on styles and trends"— and a subtext of much of the editorial decision-making: *Blueprint* favored a particular set of recognizable stylistic types in architecture and design that might be characterized as postmodern, Japanese, high-tech, and minimalist. Thirdly, through its profiles of celebrity designers, its closely observed accounts of design events, and its tracking of the activities of design personalities through the "Sour Grapes" gossip column, *Blueprint* painted a wry, but mostly admiring, portrait of the designer lifestyle. Lastly, *Blueprint* engaged with style through its lively brand- and designer-name-bespangled prose and through its visual appearance—its art direction, its stylized photography, the kinds of advertising it solicited, and its self-consciously oversized format. Through its tone and appearance, *Blueprint* clearly aspired to be assimilated as another cult object into the very designer lifestyle it portrayed.

3.6
April 1988 issue of *Blueprint*, featuring graphic designer Neville Brody, photographed by Nick Knight, as the cover star. Courtesy of Design Curial.

a) "What is there to say about clothes?"[66]

Blueprint covered fashion between 1983 and 1989, but always tentatively. A self-questioning editorial in the June 1987 issue described a difficulty in talking about fashion: "What is there to say about clothes beyond mere description?"[67] In a feature on Katherine Hamnett's political slogan T-shirts, *Blueprint* made a provisional attempt to address the connection between style and politics. Hamnett was quoted as saying: "I've managed to make ecology fashionable." Clothes, she said, convey unwritten codes that are more effective than overt statements at telling us who we are: "There are messages in clothing which are non-verbal, but which express the kind of person you are. You choose your clothes, but your subconscious picks something because it represents a lifestyle—values, ideals, tribal identifications—and expresses who you are as well as who you would like to be."[68] In another article, "The Meaning of Clothes," Peter York, bylined as "style-monger extraordinaire," reported on four London fashion stores, "the worlds they represented and their milieu."[69] York did not actually address meaning; instead he focused on spotting details, tracing references, naming clientele, and devising vivid linguistic labels such as "theatrical actor-gentish" for Scott Crolla's suits, or "Tom of Finland meets Cobra Woman" for Anthony Price's tailoring.[70]

In his editorial to the April 1984 issue, Sudjic addressed the subject of how styles change with the questions: "What triggers off those curious and seemingly tiny changes in sensibility that suddenly open up from invisible fissures and produce earthquakes in the taste landscape? Why do narrow tapered jeans look absurdly antediluvian one moment, and completely the business the next?" The same issue included an article about Tommy Roberts, co-founder of Mr Freedom, the 1970s King's Road clothing and furniture boutique, whose latest venture was Practical Styling, a furniture store in the basement of the Centre Point building in London's West End with an eclectic selection of wares. Sudjic was interested in Roberts's ability to "know what's going to happen next," intuit the stylistic whim of the moment, and take risks. He saw Roberts as a kind of cultural barometer, with an ability to predict new trends—a role that Sudjic hoped the magazine would also perform.

b) "The joy of matt black"[71]

Among the architects Sudjic and Murray endorsed in the pages of the magazine were Richard Rogers, Norman Foster, and James Stirling, proponents of a late modernist services-as-structure mode of building typified by Rogers's Pompidou Centre in Paris. Sudjic and Murray each wrote books about these architects, and in 1986 they curated an exhibition at the Royal Academy on the trio. Sudjic eulogized Foster's Hong Kong and Shanghai Bank building as "nothing less than the reinvention of the skyscraper."[72] He was particularly struck by the high-tech innovations found in every aspect of the building, from its computer-programmed motorized sun scoops which delivered sunshine to the atrium all year round, to the "elegant brass dowels" used in place of "conventionally shaped keys."[73]

In the field of interiors, *Blueprint's* editors favored both a high-tech, industrial, matt black and aluminum look and a Japanese-infused minimal aesthetic. In a report on Jiricna/ Kerr Associate's interior for the Legends nightclub, *Blueprint* gushed over a "strictly disciplined colour scheme" with "polished plaster for the walls, black for the ceiling to show off the elaborate lasers, and more black for the upholstery with polished chrome and steel everywhere else, which reflects the customers and provides a touch of glitter."[74] *Blueprint* was equally enthusiastic about the puritanical minimalism of Pawson and Silverstrin's "dazzling" interior for the Wakaba sushi restaurant in Swiss Cottage, with its "all-white walls, unadorned by any extraneous detail," and the "extreme economy and elegance of means" by which the designers had subdivided a private area of the restaurant with a five-foot-high screen.[75]

When it came to product design, it was high-tech gadgets such as the NEC fax machine, the mobile telephone, and the matt black Braun ET 22 calculator that intrigued *Blueprint's* editors.[76] Sudjic described the last, with reverence, as "the supreme cult object in the sense that it becomes a constant presence. It will slip into a pocket, or fit in the hand, and inevitably it begins to affect its owner's mannerisms and the image that he projects to the world."[77] He noted the calculator's "ultra-precise mouldings," its "shiny control buttons, bright as Smarties," and its chiselled ribbing." He knew that "it isn't the real technocrats who have made such a fetish out of the ET 22 but the design groupies with an eye for its looks," and that, "left suggestively on your desk, the calculator starts transmitting all kinds of flattering signals."[78]

Among the personalities and institutions that most consistently came under attack from Sudjic and Murray were Prince Charles, who had spearheaded a public campaign against late modernist architecture of the kind *Blueprint* liked best, and the Design Council, for its increasing isolation from the concerns of the design industry, its "fogeyishness" and lack of style. *Blueprint* repeatedly drew a distinction between the "softness" and "cosiness" of old-guard design values—its figureheads were depicted as "herbivores"—with the "sharpness," "hard-headedness," and "carnivorous" nature of modern design and their own brand of modern design commentary.[79] Among the most carnivorous of the magazine's ongoing anti-Design Council commentaries was an editorial leader titled "Abolish the Design Council," in which the magazine called for the Government to dissolve its failed experiment which, in their view, was "now little more than a vulgar gift shop and sandwich bar, to disperse its teeming hordes of leaderless bureaucrats, and to set about the real task of putting design to work for this country."[80] The Design Council fought back by refusing to sell *Blueprint* in its bookstore, and some readers, including designers Nick Butler and Kenneth Grange, defended the Council with a letter that began "Your ill-informed attacks on The Design Council, culminating in the leader in your latest issue recommending the Council's abolition may do something to establish the notoriety which appears to be your main ambition, but will do nothing to advance the cause of design and designers in this country."[81]

To some extent, *Blueprint*'s provocations were a simply a case of a new generation of critics cutting their teeth. Sudjic was in his twenties at the time and, as he later remarked, "every generation makes its reputation by trashing its predecessors ... by nailing people they think are uninteresting and promoting those that are interesting."[82] But Sudjic clearly had greater writerly ambition than to be the kind of critic who was a "sycophantic handmaiden for the doers and professionals"; he wanted to interpret what was unsaid.[83] Through championing Rogers, Stirling, and Foster, and lambasting the Design Council, he explored what it would be to stake a position, yet he did not seem comfortable with use of the first-person address, critical takedowns, and engaging in intellectual sparring matches with his peers. Rather, he commissioned other writers to do these things, preferring instead to develop a subtle and detail-rich writing style and to use the satirical technique of holding a mirror to the excesses of 1980s design culture.

c) *Blueprint* as star-maker

Sudjic's method was linguistic, but the magazine reinforced the mirroring strategy through its cover imagery. *Blueprint* was unique among other British design magazines of the period for featuring portrait photographs of designers rather than products or buildings on its covers. Several of these cover images, elaborately styled by the photographer Phil Sayer, incorporated mirrors, which provide an additional reflected image of their subjects, and also suggest that the reader might catch a glimpse of themselves in what they saw and read.

Blueprint began this strategy of putting the designer front and center on the full-color A3-sized magazine, and conferring celebrity status upon designers, with its very first cover, which featured a three-quarter-height photograph of Eva Jiricna. The forty-four-year-old Czech-born interior designer had recently caught the attention of London's architecture and design media for her dramatically minimalist interiors for the upscale restaurant Caprice, the Joseph fashion stores, and designer Joseph Ettedgui's Sloane Street flat.

For the *Blueprint* cover shot she was captured emerging through a mirrored door, her hands clasped around an electric lamp, echoing images of Florence Nightingale with her oil lamp. Her face and hair shone in the dramatic lighting, and she smiled a closed-mouthed smile, as if bemused by the attention of the photographer. "Architects, blinking mole-like in the sudden glare of unfamiliar exposure, were nonplussed at first," Sudjic later remarked of their increasing celebrity status.[84] Jiricna was literally framed within a *mise en scène* of her own making, the bathroom of her Belsize Park flat, which featured bright green industrial rubber dot floor material as wall covering and a porthole, and was accented with a pair of nautical buoys hanging from an S-hook. She wore a white round-necked top with an oversized chain necklace and black belted trousers for the shoot, which readers learned in the profile, written by Maurice Cooper, was her "signature look."

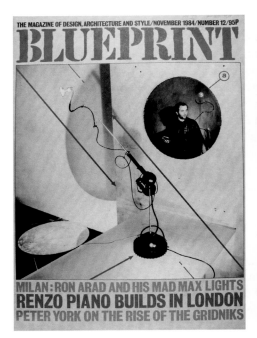

3.7
Cover of *Blueprint*, November 1984, showing the use of mirrors in Phil Sayer's portrait of designer Ron Arad. Courtesy of Design Curial.

3.8
First issue of *Blueprint*, October 1983, featuring Eva Jiricna, the Czech-born interior designer, photographed by Phil Sayer. Courtesy of Design Curial.

The image captured Jiricna's sensibility as a designer, and because it was taken in her own home, it also provided the viewer with a sense of privileged access to the designer's life beyond the studio. Cooper's profile provided more detail about his perception of Jiricna's personality:

> Home now is a flat in Belsize Park, the architects' ghetto. You can't help noticing that most of the furniture, from the life jackets on the sofa to the bulkhead lights strung up on yacht hawser, seem to have come from a ship's chandlery. The living room feels like a swimming pool, with a vivid green rim and deep blue carpet. The dining table is perforated black metal and folds up out of sight. And the kitchen is more galley than anything else. It's a tough uncompromising place to live, which is just as she wants it. Every item in the flat has been chosen with measured care, pondered over and debated, just as the rest of her interiors.[85]

The magazine's first editorial statement declared its intention to take a personality-centric approach: "we will be profiling the tastemakers and talking to designers and architects in a way which, we believe, is not currently being done by either the professional or the lay media."[86] Few other architecture and design publications of the period ran profiles of practitioners. *Blueprint*, drawing from an approach found in music and society magazines, wanted to capture the designer as a personality, including details of where they ate and what they wore.[87]

In an era fascinated with the construction of identity, the journalistic device of the interview assumed new significance. With its intimate, lengthy, and often unedited interviews with celebrities, the American magazine *Interview* typified the genre. As Paul Atkinson and David Silverman have observed, "The interview, with its implied invitations toward self-revelation, is a pervasive device for the production of selves, biographies, and experiences. It furnishes the viewer/reader/hearer with the promise of privileged— however fleeting—glimpses into the private domain of the speaker."[88]

Blueprint's profiles of prominent designers were based on interviews, with some of the most incisive pieces written by Janet Abrams, who clearly relished the potential of the format for intense debate with the big minds in design. She considered that her role as a journalist, far from merely reporting the facts, was to figure as a co-protagonist in the story. A profile of architect Peter Eisenman began: "'Are you going to do a number on me?' Peter Eisenman inquires when I phone to arrange this interview. One of the distinguishing characteristics of Ivy League architect-academics is to veil their rapturous delight in publicity with feigned outrage at the mere prospect; Eisenman is perhaps the archetype of the genus."[89]

Until *Blueprint* began to prioritize the interview-based profile as a journalistic format, most designers had stood well behind their work, keeping their private lives separate from the more public and carefully constructed environments of their studios. Increasingly writers asked for access to designers' homes and observed them in their daily lives, giving rise to a new kind of pseudo-psychoanalytic character analysis in design journalism.

Sudjic was a keen observer of identity himself. In a 2006 article about his first visit to his parents' hometown of Belgrade in twenty-five years, he traced the roots of his own need to decode identity as a response to his immigrant parents' and his own uncertain status in relation to British culture and entrenched class divisions. He wrote: "How identity is manufactured has always interested me from the first time that I began to wonder why money in Yugoslavia was in the form of banknotes embellished with portraits of heroic power station workers and apple cheeked peasants, and in Britain money is signified by men with whiskers and big wigs. These are the clues that you need to decode in order to get a grip on exactly who you are."[90]

Sudjic attempted to answer the question "Are you what you own?" in *Cult Objects: The Complete Guide to Having It All*, a 1985 catalog of the accouterments of the designer lifestyle.[91] The same year, his peer Stephen Bayley's exhibition and catalog *The Good Design Guide: 100 Best Ever Products* seemed to be motivated by the same question. The *Guide* featured such status-conferring design icons as 501 jeans, Oxford shirt, Panama hat, Zippo lighter, Oyster Rolex, Raybans, Bass Weejun shoes, K100 motorbike, and a Porsche pipe.[92] As evidenced by numerous articles of the period, Bayley owned and used nearly all of these products. A 1986 advertisement for Herbert Johnson Panama hats in *London Portrait Magazine* proudly referenced its selection as one of the "100 best designed products" for a Boilerhouse Project exhibition.

Sudjic's book and Bayley's catalog were visual anatomies of the well-appointed urban male *flâneur* who oriented himself in 1980s London as either a producer or a consumer of designed objects, and often both. The taxonomy as a format suited 1980s design journalism, which was coming to terms with the new emphasis on design as a lifestyle choice. By classifying the visual attributes of the designer lifestyle, Sudjic and Bayley were establishing their vocabularies, charting their territories by naming and identifying, but not analyzing, and certainly not critiquing, the unspoken status anxiety that underlay their projects. The tone of each writer differed. Bayley was earnest. He hoped his *Guide* would provide "exemplars for imitation in the future" and a "stimulus for creativity."[93] His descriptions of each object were assured to the point of dogmatism: "The Rolex Oyster has become the archetype of the wristwatch, an unimprovable classic of design that has often been imitated but never surpassed."[94] Sudjic was more ambivalent: in the case of the Mont Blanc, for example, he both lovingly described the pen's attributes and exposed the "largely spurious" nature of its "archaic" styling, concluding: "It is an upstart pretender, a fountain pen born of the Biro and Pentel era, and manufactured in Hamburg by a subsidiary of the Dunhill tobacco empire."[95] But neither author questioned the "cult" of designed objects, nor the confusion about "where to draw the line between who we are and what we have," as Hebdige put it in a *Blueprint* essay on late-1980s décor magazines.[96]

d) Using prose to pose

Sudjic was fascinated by the manifestations of style and lifestyle he noted in 1980s London—the objects, clothing, and behaviors that signaled knowingness on the part of the adopter. He was a subtle observer of stylistic codes; moreover, he possessed the linguistic panache to transmute them into prose. Due to *Blueprint*'s budgetary constraints, Sudjic had to write most of the copy in the early issues, so it was largely his writing that defined the prose style of the magazine.

Sudjic was interested in the writing of several of his peers but says his biggest influence was Reyner Banham, via his weekly columns about the complexion of everyday design products in *New Society*. As editor of *Blueprint*, Sudjic appreciated that a new approach to writing about design required a new type of vivid, entertaining prose, which he has referred to as "gloss."[97]

The style-conscious, brand-labeling, and semifictional approach to design writing exemplified by Sudjic's prose anticipated to some extent, and coexisted with, the fiction writing of American brat-pack novelists Jay McInerney and Brett Easton Ellis, whose characters in *Bright Lights, Big City* and *American Psycho*, for example, were defined by the products and status-conferring totems they surrounded themselves with. It also belonged to the literary lineage of New Journalism, popularized in the 1970s by American authors such as Gay Talese and Tom Wolfe.

Wolfe was an important literary touchstone for other British design commentators of the period, as well as Sudjic. They had read his nonfiction writing from the 1960s and, more recently, his account of what he saw as the repressive tyranny of architectural modernism in *From Bauhaus to Our House*, published in the UK in 1981, as well as his *Harper's* and *Esquire* articles collected in the 1976 book *Mauve Gloves & Madmen, Clutter & Vine*. Bayley made frequent reference to Wolfe in his writing, and corresponded with him about his work. In 1983, for example, Wolfe wrote to Bayley about one of his exhibitions, saying: "I've been following with great amusement the furor over Taste. It's marvelous."[98]

When Wolfe came to England to give a lecture at the University of Kent on a damp November evening in 1983, the *Times Literary Supplement* reported that all the "acolytes of style, hot off British Rail, are there to hear him: Faber & Faber (the funding fathers), Harpers & Queen, the RCA, and the Boilerhouse Project."[99] Meanwhile, *Blueprint* registered "Sloane chronicler Peter York, in the second row, taking copious notes, like a keen young undergraduate."[100] Wolfe's lecture, "The Trend Who Walks Like a Man," was about the "vast sociological experiment" that is New York's SoHo—"the lifestyle of the 100,000 registered artists clustered in the lofts and rookeries."[101] *Blueprint* recounted: "we meet the artists, the dealers, and their girlfriends, we discover how they live and—in great detail—what they wear."[102]

The techniques of New Journalism included immersing readers in a dramatic scene through in-medias-res beginnings, and in the private world of the subjects' minds through the third-person point of view, use of the first- or second-person address, the historical present tense, long sections of dialogue, and the deployment of narrative prose saturated with detail and exaggerated metaphor. In characterizing New Journalism, Wolfe also emphasized the importance of assimilating lifestyle—the "recording of everyday gestures, habits, manners, customs, styles of furniture, clothing, decoration, styles of traveling, eating, keeping house ... the accumulated details that symbolize status as a way to immerse and absorb the reader."[103] Visually it looked different from other journalism, too, characterized by a playful and abundant use of dashes, dots, and exclamation points, which Wolfe said helped him "give the illusion not only of a person talking, but of a person thinking."[104]

A self-described "prose stylist," York, too, was enamored of Wolfe's formal panache, and his ability to recognize "the entire pattern of behavior and possessions through which people express their position in the world or what they think it is or what they hope it to be."[105] The fact that Wolfe seemed to suspend his judgment between love and hate of his subjects, and could move so nimbly between the worlds he reported on without being weighed down by a fixed viewpoint, was also appealing to York. In an article about Wolfe for *Harpers & Queen*, York quoted *Harper's* editor Lewis Lapham saying of Wolfe, whom he regularly commissioned, "He has a view of US society not shrouded by cant, he sees it in terms of money, sex and class, he's free from ideological arguments."[106]

There was a macho quality to New Journalism that may have represented part of its allure to writers like Sudjic, Bayley, and York. York appreciated Wolfe's continued liberating influence on English journalists, both directly through his own writing, and indirectly through a generation of rock critics and color supplement writers whom he had influenced in the 1960s. York, like Sudjic, cited Reyner Banham's "visually-oriented pop/sociological" brand of writing in *New Society* as a key example.[107] More than this, though, York revered Wolfe's notoriety—the fact that he was the "first journalist other journalists wrote about," and that he provided writers like himself with a role model, and journalism itself with a new celebrity status. In his explanation of English journalists' "hyper-awareness" of Wolfe, York wrote: "Wolfe was a celebrity to other journalists all right. And he had done more even than set an example in subject-matter and style and celebrification, he was providing the building blocks of a rationale that said journalists—journalists on the lowbrow papers and slick magazines and the specialist press—could be in the fast lane of Modern Culture, pushing deadbeat novelists off the road."[108]

Blueprint's gossip column, titled "Sour Grapes," provides the most emphatic evidence of Sudjic's negotiation of the "fast lane of Modern Culture," and his urbane and often acerbic writerly voice. "Sour Grapes" was based on the snarky digs, bold-type names, and third-person anonymity of gossip columns in society magazines and "Pseud's Corner" in *Private Eye*, and on the tone of New York satirical magazine *Spy*, which lambasted the vices and follies of American media personalities like Donald Trump. "Sour Grapes" recorded, and created, the controversies of the design scene, and provided a soap-opera-like running commentary on the lifestyles of the designers featured in *Blueprint*. The column reflected Sudjic's preoccupations and prejudices—missteps made by Boilerhouse curator Stephen Bayley, and disdain for *bêtes noires* such as his ex-employer the Design Council's out-of-touch perspective on the design profession, and Prince Charles's widely publicized campaigns for traditional architecture. It also provided closely observed reports on the appointments, firings, achievements, and mistakes of London's architecture and design writers and editors, demonstrating another way in which the magazine was self-aware of its role within the larger context of an evolving design media industry.

Blueprint played a key role in taste-making politics of the 1980s. Sudjic's values were evident in his selection of subject matter, but also through the judgmental language he used. Terms like "vulgar," "upmarket," "downmarket," and "brash" littered the magazine's pages. The young architect Nigel Coates was described as being "not quite one of us," and Sudjic remarked of the crowd in Milan that "everybody who was anybody was there."[109]

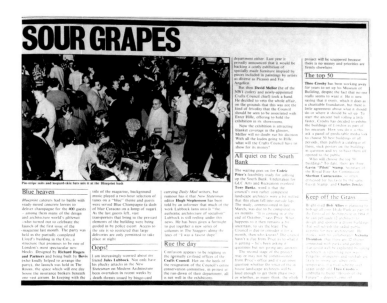

3.9
"Sour Grapes" feature in *Blueprint*, showing a photograph of the magazine's first issue launch party, held at the partially completed Lloyd's Building designed by Richard Rogers and Partners, in October 1983. Courtesy of Design Curial.

Coverage of the annual Milan Furniture Fair provides a telling example of *Blueprint's*, and primarily Sudjic's, fascination with the designer lifestyle. For the occasion he translated "Sour Grapes" to "Grappa Acido," and printed photos of design celebrities at the various parties. Readers were told: "Those not invited [to the Tecno party] are allowed the food but not the present. Last year everyone got a Swatch watch—the Swiss answer to the Japanese domination of the watch industry. This year they gave radios the size of a credit card."[110] At the Memphis party, to which thousands flocked but only a handful gained admittance, "Ettore Sottsass arrived after a couple of hours and moved regally through the throng, kissing and shaking hands with the adoring fans as he passed."[111] Sudjic always noted designers' attire (furniture designer Ron Arad was "wearing a bowler hat," while architect and editor of *Domus* Alessandro Mendini "sports designer jeans") and their jet-set lifestyles (Terence Conran "popped into the Fair on his way back from New York to London," and Norman Foster "flew over in his jet for the occasion.")[112] In another piece about the Fair, "Milan: The Party Is Over," Sudjic employed the scenic immersion techniques of New Journalism to transport his readers viscerally into the Milan Furniture Fair experience: "It's just getting dark as we step out of Vico Magistretti's party at the Cassina showroom into a sticky Milanese dusk full of sirens and orange trams, when a glistening face detaches itself from the ravening hordes of Paolos and Tomassos gulping Cassina's white wine and shoveling down Cassina's caviar sandwiches."[113] He described the excessiveness of the spread at another party with a characteristic mixture of relish and repulsion: "there are relays of white-gloved waiters, decked with chains of office, dispensing champagne, mountains of langoustines, baby octopuses, risotto and blueberries to brawling crowds of elegantly tanned ladies wearing great chunks of brass around their necks and wrists."[114] He deconstructed the social hierarchies of Milan, "the Design World Headquarters," allowing *Blueprint* readers who were there to feel validated, and those who weren't to share vicariously in its business machinations, social pleasures, and sartorial details.

In 1989 York drew attention to the role of satire in the wider popular assessment of the design boom. He noted that "design has itself started to turn up in plays and films of a satirical or left wing kind as a metaphor for whatever their writers see as dishonest or manipulative for the 1980s."[115] York's own satirical portraits of designers could be found among his essays in the magazine *Harpers & Queen* during the late 1970s, collected in the book *Style Wars*, and in those he contributed to magazines such as *Vanity Fair* in the early 1980s, collected in the 1984 book *Modern Times*. When *Modern Times* was published, *Blueprint* excerpted one of its essays, "Chic Graphique," which sends up the lifestyle and trappings of what York called "the Graphic," a social archetype who epitomizes the early-1980s graphic designer, but also other design aficionados, who are clearly meant to include the *Blueprint* readership and its contributors.

In his review of York's essay collection, Sudjic portrayed York as an urban entomologist, "wielding his butterfly net over Homo Covent Gardeniensis," who saw the potential in design as satirical fodder. "He dissects the foibles of the breed with merciless accuracy,"

Sudjic wrote, approvingly, of York's pinioning of the graphic designer's home furnishings (exposed structures, nylon door handles, white tiles, teaspoons in an old Keiller marmalade pot), grooming habits (short, even-length, all-over beard/mustache), clothes (Paul Smith cashmere red scarf, raspberry-colored jelly-framed glasses, denim shirt with no tie), accessories (Mont Blanc or Lamy pens, metal mesh or one-piece molded polythene carrying cases in primary colors), heroes (Milton Glaser, Gropius, Bruce Weber), and horrors (*Interiors* magazine, Laura Ashley, herbaceous borders).[116]

York's tone was arch, yet he did not actually puncture the designer lifestyle bubble, and he certainly did not draw attention to its complicity with Thatcherite politics or its unsustainable production and consumption practices. The eye that saw all those details was essentially detached and amused, not angry. York did not consider what he did as critique, in the sense that criticism might have a moral or political purpose. He later observed: "a critique has to come from a fixed position, doesn't it? I saw myself as giving a bit of fun along the way. It wasn't my concern whether the nation got a good deal from serious designers, or whatever."[117]

York wrote colloquially, rhythmically, in the present tense, addressed the second person, and used allusive vocabulary and lists of references to be appreciated by those in the know. But for York, even though he loved writing in the sense that it was a "performance" through which he could "show off," journalism was a hobby. He was primarily a marketing consultant and co-founder of the management consultancy SRU Ltd, who was adept at characterizing "tribes" firstly as potential markets and only secondly as topics for his journalism. York quickly discerned that the "designer lifestyle" he saw taking hold in 1980s London was both a new market for his clients and good subject matter for his writing:

> I came from outside the stockade and on the face of it with unkindly intentions. The things people say and wear, things like fell walker shoes, were funny but it was also important. And if you have these factors on an upward trajectory, it's got to be something to write about. The design classes, that movement, we were seeing from the world of the word, of which literary novelists would be part, and a whole swathe of other kinds of academics. You just had this gorgeous material. There are times when particular things are hot. I recognized it with every instinct I had.[118]

e) *Blueprint*'s "hardcore" image[119]

Blueprint continued its engagement with style through its art direction. Simon Esterson, *Blueprint*'s art director, and certainly a "Graphic" by York's measure, developed a bold and distinctive design for the magazine that used all-caps blocky headlines, an architectural compositional system of thick rules and text boxes, coupled with oversized photographs that took full advantage of the A3 size of the pages. Mainstream British graphic design of the 1980s was going through what Esterson terms "a classical, centred, woodcutty phase," typified by packaging design by Michael Peters and Trickett and Webb. Esterson responded to the experimental work of other designers of the period, like David King and Neville Brody, who were rediscovering Russian Constructivism and using its visual energy to infuse their own graphics for political movements such as Rock Against Racism and Red Wedge. In particular, Esterson channeled the tough, urban graphic style of the New York architecture journal *Skyline*, which was designed by Michael Bierut, then at Massimo Vignelli's office.[120]

The tabloid newspaper format was chosen to emphasize the intentionally ephemeral nature of the project.[121] "We wanted it to last for only ten issues and then die," said Sudjic. "We deliberately chose the *Blueprint* format as a disposable one." He and Murray thought the awkwardness of *Blueprint*'s shape would prevent it from being filed with other magazines in a design studio library, therefore it would be thrown away. As design historian Liz Farrelly has observed, this emphasis on the magazine's ephemerality is in fact part of the somewhat disingenuous myth-making that surrounds its inception, since an advertisement for back issues appeared in issue 8, June 1984, acknowledging the fact that architects and designers would want to keep and file this stylish-looking object.[122]

Style was also an integral part of *Blueprint*'s business model. As Sudjic characterized it in 1989, rather than "bending over backwards to write about advertiser X's chairs or advertiser B's office furniture systems," the magazine played "hard to get," making it seem "like quite the place to advertise." He reflected that the style of the magazine was what helped to attract advertisers: "It is current and it is presenting things in a stylish kind of way, and the advertisers see that reflecting on their products and they want to be in it."[123] Furthermore, Esterson would often redesign the submitted advertisement artwork, ostensibly because of *Blueprint*'s unusually large format, but mainly in the sense of tidying it up, making it more consistent with the visual tone of the magazine.

Sober reflections

Although Sudjic did not leave the magazine until 1993, toward the end of the 1980s he became more self-questioning, alluding in his writing to the ways in which the magazine's creators might have been implicated in the creation of an inflated view of design. To mark *Blueprint*'s five-year anniversary in 1988, its editors devoted a special issue to surveying the "design decade." Sudjic's editorial leader reflected back on *Blueprint*'s role in chronicling Britain's design boom. Throughout the 1980s the design industry expanded exponentially, as a service to business. Sudjic observed how close big-business design and Thatcherism had become by this time—how the "once essentially liberal profession of design has accommodated itself so readily to the new orthodoxies."[124] At a practical level, Thatcher's government supported design's entrepreneurial growth. But ideologically, the similarities between Conservative Party politics and design's applications were more striking and therefore presumably discomfiting to someone like Sudjic, whose architectural training rested on liberal and idealist philosophies. Sudjic viewed the situation as a detached observer, however, rather than an implicated player. He wrote: "The present-day business of design, with its stock market listings, its takeovers and its tycoons, might be taken as a metaphor for the Thatcher years. Indeed design is in danger of becoming so closely associated with Mrs. Thatcher's brand of radical conservatism that it may yet find itself in real difficulties in a post Reagan and Thatcher era."[125] In doing so, he framed the Thatcherite entrepreneurial spirit as just another style, that had been unquestioningly adopted by design culture during the mid-1980s and which would eventually be disposed of, when issues such as ecology returned to favor with the return swing of the fashion pendulum. In another end-of-decade editorial Sudjic pondered how design might redefine itself in the coming decade with the prospect of a Labour government seeming like a real possibility: "Will it seek to ally itself with the green movement and social responsibility …?"[126]

In 1989, the same year as a collection of Phil Sayer's *Blueprint* cover-star portraits was installed on the wall of the Blueprint Café at the newly opened Design Museum, Sudjic wrote the book *Cult Heroes: How to Be Famous for More Than Fifteen Minutes*.[127] One of the book's chapters charts the rise of the architect and designer to celebrity status, a phenomenon that *Blueprint* had encouraged, yet one that Sudjic again regarded with characteristic detachment: "fame has become the most valuable, the most sought after, and the most perishable of commodities."[128] He wrote of the 1980s as a decade "that has become addicted to the cult of personality"—a curiously passive turn of phrase, which deflected responsibility away from his own editorial decision-making and toward the culture at large.[129] Elsewhere he wrote of the "double-edged nature of media attention, and the way design is trivialized when turned into fodder for the consumption of an image-obsessed society."[130] In *Cult Heroes* Sudjic castigated the media's short attention span, again keeping his own involvement at arm's length through use of the passive voice: "Designers may achieve brief periods of fame and fortune, but all too soon find themselves discarded … their work exhausted of meaning and content."[131] Verbally, Sudjic could be more candid about his culpability in boosting designers to star status. In 1989, questioned about this, he said: "maybe we've helped [the whole star thing] along a bit in design, helped invent a few design stars."[132] In an editorial discussing the folding of the Milan design collective Memphis, Sudjic wrote of the design media's role in flattening the complexity of much design: "The real lessons of the Memphis movement, however, will be the double-edged nature of media attention, and the way design is trivialized when turned into fodder for the consumption of an image-obsessed society."[133]

Sudjic's accomplishment as a prose stylist, his ironically detached stance as a reporter, and his entertainment-based approach to editing were perfectly attuned to the exuberance and fetishism of the dominant strain of 1980s design culture. Whether he liked it or not, by the late 1980s *Blueprint* was inextricably enmeshed—economically through its advertising revenue, and ideologically through its editorial choices—with the values of enterprise culture design. The magazine would have to be significantly retooled to deal with the sober topics of design for the public realm, social responsibility, and environmentalism that Sudjic perceived to be on the horizon.

3.10
Photograph of the "Art and Industry: A Century of
Design in the Products You Use" exhibition held
at The Boilerhouse Project in 1981. Courtesy of
The Design Museum London.

3.11
Entrance to The Boilerhouse Project, showing
the white tiles and clinical, hospital aesthetic.
Courtesy of ERCO GmbH.

DESIGNER CELEBRITIES AND "MONSTROUS, BRINDLED, HYBRID" CONSUMERS

PART TWO: THE BOILERHOUSE PROJECT AND THE DECONTAMINATION OF TASTE

"The sterilized operating room of the white cube"[134]

When home furnishings entrepreneur Terence Conran profitably floated his Habitat chain on the stock market in 1981, he used some of the proceeds to set up The Conran Foundation, a charity dedicated to improving public appreciation of good industrial design. Conran had initially wanted to publish a magazine, with Sudjic as its editor, but when the Victoria & Albert Museum offered him its disused basement boiler rooms, he decided that an exhibition space would be just as good an outlet for his aims. He selected as its director the Kent University art history lecturer, and author of *In Good Shape: Style in Industrial Products 1900–1960*, Stephen Bayley, to whom he had been introduced by Paul Reilly, director of the Design Council.[135]

The Boilerhouse and *Blueprint* magazine coexisted, if not in harmony exactly, then certainly along close parallel lines. *Blueprint* assiduously reviewed each of The Boilerhouse exhibitions, a habit established in the first issue with James Woudhuysen's review of its "Taste" exhibition.[136] Some of *Blueprint*'s writers—Jonathan Glancey, for example—curated shows for Bayley. Furthermore, the exhibitions themselves were like three-dimensional magazine articles, and appear to have been written into existence like texts, an impression underscored by the use of Bayley's own handwriting for the lengthy wall texts and captions in the "Taste" exhibition.[137]

Although at first the 500 square meters of disused and flooded space given to Conran and Bayley looked to Bayley like "a fetid bunker," it was quickly renovated by Conran Associates. They covered the walls and floor with bright white tiles, and installed Erco spotlighting in the ceiling, thus creating a pure chamber of exhibition space, isolated from the outside world, and seemingly following the prescription for modernist art display that art historian Brian O'Doherty had charted in his 1976 *Artforum* essays about the ideological construction of a "white cube" context for art. O'Doherty wrote: "The outside world must not come in, so windows are usually sealed off. Walls are painted white. The ceiling becomes the source of light. ... The art is free, as the saying used to go, 'to take on its own life.'" In O'Doherty's conception, art had been inoculated against the real life of the world and the contingencies of time and change, through its containment in "the sterilized operating room of the white cube."[138]

The Boilerhouse Project's version of the white cube was frequently described as clinical. A *New Musical Express* journalist remarked that "It looks like a private hospital," a loaded simile, since the Thatcher government had begun major reforms of the NHS with the aim of pushing many toward private hospitals.[139] Peter York referred to it as "Emergency Ward 10."[140] The conservative art critic Brian Sewell in the *Tatler* called it "a subterranean installation so aesthetically hygienic that it seemed to have been sanitized for our protection."[141] These descriptions recall Jules Lubbock's evocation of the Saatchi Gallery as "30,000 square feet of whitewashed and windowless gallery." Lubbock extrapolated, just as Stephen Spender had in the 1950s, that "modernists are obsessed with hygiene. It is the Hoover and deodorant style. ... Mrs Thatcher doesn't smell. Not a whiff of a phero-mone escapes her armpits."[142] Through his exhibiting practice, Bayley can be seen to have functioned like the hygienic modernist Lubbock had conjured, seeking to cleanse the clut-tered and dirty popular notions of taste, and the messiness of postmodernism, with his own organized and sanitary vision.

The Boilerhouse's white tiles also formed a graph-paper-like backdrop, which meant objects were always seen in their pure, drawing-board state, uncontaminated by use. Furthermore, the clinically white, frictionless, and disorientating stage set created by Conran Associates could be said to manifest a contemporary condition that Jean Baudrillard had termed "simulacrum." According to Baudrillard, signs had become increasingly dis-connected from the things they referred to, until by the 1980s people inhabited a hyperreal universe made up only of signs, surfaces, and images circulating with no connection to any real world outside themselves.[143] Baudrillard was fascinated by theme parks, political campaigns, television shows, conferences like IDCA, and museums, arguing that these simulations hide not reality, but the disappearance of reality.[144]

The Boilerhouse exhibitions followed a fast-paced schedule of around five per year, each with its own catalog, and were researched, assembled, and designed with the rapid-ity of magazine features. Indeed, in characterizing his approach to curation during this period, Bayley said: "I was doing journalism in three dimensions. So I would just set up an argument, a debate, and flesh it out with objects."[145] The exhibitions ranged from explora-tions of the values and mechanisms of design, such as "Taste" (1983) and "Art & Industry" (1981) to showcases of archetypes or trends such as "Robots" (1984) and "Post-modern Colour" (1984) and blatant celebrations of commercial brands that Bayley deemed stylish, including Sony (1982) and Coca-Cola (1986). "The Boilerhouse has at times seemed but the upmarket public relations side of selected manufacturers and designers, and has not produced any magical method for dispelling the fog of suspicion between designers, artists, industry, public," wrote Marina Vaizey in the Sunday Times. "Yet," she continued, "if this exercise is flawed, a real sense of debate has nevertheless been engendered, and a new public is being reached."[146]

A project, not a museum

Bayley was keen to distinguish his activity at The Boilerhouse from museum curation, per-haps due to his negligible experience as a curator, but also because of The Boilerhouse's mission to be "an abrasive stimulus to the public."[147] He claimed that he "always fought against preposterous conceits and vanities of the museum establishment and their art-historical indulgences."[148] Exhibition-making, in Bayley's view, was "something more" than museum curation—"It has to be more like theatre." To create this sense of theatricality he used attention-grabbing exhibition design, such as John Pawson's extreme minimalist design for the 1984 "Handtools" exhibition, which used long, low, black wedges to display the objects, so that visitors had to bend down to see them. Bayley also manipulated the media skillfully, encouraging them to report on any controversy that arose around the exhibitions—The Boilerhouse generated 16,000 column inches of often vituperative, national and international press coverage—thus helping to increase the theatricality of what went on in his Boilerhouse stage set.

Bayley and Conran were careful not to refer to The Boilerhouse as a museum, because they disliked the moral certainties associated with museological conventions. Instead they called it a "found object with readymade industrial overtones," and the use of the provisional term "Project" in the title connected it more to the active work of a design or architectural studio than to the institutional construct of a museum.

The Boilerhouse was administratively light on its feet, with no permanent collection and no keepers; it emphasized its clean, modernist aesthetic in opposition to the Victorian galleries and antiquated display cases in the main building of the V&A; it addressed popular culture head-on; and it was privately funded and deeply enmeshed with commerce. It was also shaped by a subjective, editorializing approach. Inspired by Henry Cole's "Chamber of Horrors," which in 1852 had pilloried examples of "false" contemporary design, Bayley said he had "never been worried about putting my judgment on display."[149] The Boilerhouse, as an exhibiting apparatus, therefore, was Conran and Bayley's critique of the institutionalized methods of collecting, curating, and exhibiting design.

Nevertheless, The Boilerhouse was nurtured by its host institution, as the critic and historian Robert Hewison put it, "like a mutant strain ... within the viscera of the V&A Museum."[150] And, of course, in 1989 Conran and Bayley turned the Project into an actual museum, a cornerstone of Conran's £200 million Docklands Butler's Wharf redevelopment project. By that point Conran was more than happy to use the label "museum" to help confer cultural status upon his development, not least because, as a double-page advertisement for the project stated, it would set "the tone for retailing."[151]

It was always Conran's goal that The Boilerhouse should increase his market base and deliver more educated and eager consumers into his stores.[152] This aim was fulfilled most explicitly when Conran's own goods, or the appliances of companies he endorsed through Habitat, such as Russell Hobbs or Braun, were included in an exhibition. The larger goal, though, was to make the exhibition visitor feel as if he had good taste, and to empower him to demonstrate this discernment through buying things. This meant introducing the potential consumer to the accouterments of a modern designer lifestyle and allowing him to feel familiar in this milieu, so that next time he happened upon a Habitat catalog, the world it represented would feel recognizable and he would be ready to make informed purchases. A 1982 *Reader's Digest* article credited the "Conran style" with an ability to span "age groups, class barriers and national boundaries," and quoted a *Le Monde* piece which said: "The Habitat style is a phenomenon of our times, so well-defined that no one who buys there needs a decorator. It is not just a style but a lifestyle."[153]

As Bayley remembered it, even though Conran never asked for his products to be included in exhibitions, the business arrangement was such that "'I'll give you a million shares, but part of the design education has to be about teaching people about the Conran Way. The more you promote awareness of design the more they'll go to Habitat and the more money I'll give you.'" Bayley recollected: "It was meant to be a glorious circle."[154]

Glorious indeed, for Conran, for the V&A, which benefited from the increase in visitors and was able to test the market for its Twentieth-Century Gallery, which opened in 1989, for the numerous design and architecture magazines spawned around this time—especially *Blueprint*, for which The Boilerhouse provided so much material for controversial quotation—and glorious for Bayley, whose career as a mediagenic style guru was launched so effectively through his role as director. But this circular flow of culture and capital was more problematic for critics who were skeptical about the actual value of the Habitat lifestyle to the general public. As Judith Williamson remarked, "lifestyle and lifestyle choices makes an overlay, a thin veneer, on distinctions that are actually *class* distinctions. The idea that you can choose your place in society through the things you buy is complete nonsense."[155]

Bayley's dust-off canister of taste

In the autumn of 1983 The Boilerhouse mounted "Taste: An Exhibition about Values in Design." The exhibition, and the catalog, which anthologized authors ranging from Henry Morley and Charles Eastlake to Nikolaus Pevsner and Jules Lubbock, presented a historical narrative of the way a confluence of taste and design had been philosophized upon, constructed, and materialized through objects at different historical junctures from the eighteenth century onward. The concept of taste encompasses issues of class and social, economic, and cultural capital, thus the show was calculated to provoke the general public, but it was also meant to unsettle discourse around design criticism itself, since taste was, as Bayley observed, "among the processes we use to make judgments about design."[156]

The exhibition was designed by the graphic design firm Minale Tattersfield, who, with Bayley, developed a conceit whereby objects were displayed either on upturned galvanized steel dustbins or on white plinths, depending on Bayley's view of their taste value. The identity for the show was rendered in a three-dimensional model at the entrance, which functioned as a key to the exhibition's organizing device. The word "Taste" was spelled out with the "T" in Roman type made out of oak and resting on a white plinth to indicate the tasteful end of the spectrum; the "E," made of pink synthetic fur and resting on a dustbin, was aesthetic worlds away.

The exhibition was divided into sections: The Antique Ideal, Mass Consumption, A New Way, The Romance of the Machine, Pluralism, and Kitsch. Bayley defined this last category as "the dark side of the optimistic pluralism which characterized the post-War decades," and among the objects he used to illustrate the concept were Louis Quinze-style furniture from Maple, Waring & Gillow, a 1981 Toyota Celica instrument panel, a Harrod's Bambu dining chair, and an assortment of replica Wedgwood Coalport cottage pastille burners. One of the most publicized controversies from this show arose when Bayley put a model of the architect Terry Farrell's postmodern TV-AM Studios in the "Kitsch" section, albeit resting on both a dustbin *and* a plinth. This decision incensed Farrell, who wrote a letter of complaint and, on the second day of the show, sent members of his studio to remove the model.[157] Bayley left the plinths where they were and, in place of the model, he displayed Farrell's letter and a Polaroid he had taken of the model being carried away. This move, and Bayley's loaded descriptions of postmodern architecture such as "ham-fisted decoration, the techniques of shoplifting rather than building," also upset some architecture critics, such as Colin Amery, who, writing in the *Financial Times*, said he saw this as evidence of "how far Bayley is from understanding the new climate of Postmodern architecture."[158]

Other critics objected to use of the term "kitsch" itself. The German design critic Gert Selle, who curated an exhibition titled "Genial Design of the '80s" at the International Design Center in Berlin, also in 1983, argued that the concept of "kitsch" was anachronistic, and insufficient for understanding contemporary design: "It seems extremely doubtful whether a classification of the esthetic manifestations of contemporary product culture into *kitschy* or *non-kitschy* is a performance of insight," he wrote. Instead, Selle made a case for a more complex "social-esthetic empathy" with which to overturn the "absolute of good taste" and recognize one's own implicated position in contemporary product culture.[159]

3.12
Title wall for the "Taste" exhibition held at The Boilerhouse in 1983 and designed by Minale Tattersfield. The "good" taste serif T is on a white plinth, while the synthetic fur E is on a dustbin, setting up the binary conceit for the exhibition. Courtesy of ERCO GmbH.

Bayley's choice of a dustbin as a display device did not refer to any discourse around sustainability or a critique of built-in obsolescence, although it may have conjured recent memories of the 1979 dustmen's strike in London, when uncollected rubbish was strewn around the streets, prompting concerns over public health. *Harpers & Queen*'s Anne Engel asked: "Is the museum to become ... the show-place of the detritus of a Keep Britain Tidy campaign?," and then noted a comment in the visitors' book that simply concluded: "Rubbish."[160]

The use of new and shining galvanized steel dustbins as display devices in the clean, white-tiled gallery environment that resembled a hospital is emblematic of The Boilerhouse's attempts to define a sanitized territory for design, and thereby to repress and extinguish the sickness and pollution of everyday life. "Modernism means an overwhelming urge to tidy up. And we wanted to show what benefits tidying up could bring," Bayley said.[161]

3.13
Ornate Victorian couch by Johann Heinrich Belter, displayed in the "Taste" exhibition at the Boilerhouse in 1983, and supported on galvanized steel dustbins to indicate its tastelessness in the value system of the exhibition. Courtesy of The Design Council/ The Manchester Metropolitan University.

The anthropologist Mary Douglas has studied the symbolic nature of impurity and dirt in relation to a range of societies, writing that "reflection on dirt involves reflection on the relation of order to disorder, being to non-being, form to formlessness, life to death."[162] Her observations of the ways in which societies react to dirt point to an illuminating parallel in the ways in which curators, retailers, editors, and writers often approach designed objects: "Dirt offends against order. Eliminating it is not a negative movement, but a positive effort to organize the environment."[163]

Curating an exhibition about taste-making involved drawing evaluative distinctions between dirty, disordered, distasteful real life and the carefully selected, hygienic constructions of an idealized and exotic designer lifestyle. Tellingly, Bayley's "favourite toy," as reported in a *Times* article, was a "Falcon Safety Products Dust-Off canister of compressed air for blasting dust away."[164]

As literary theorist and sociologist Roland Barthes observed of 1950s French commercial culture, manufacturers of cleaning products needed to fix the idea of dirtiness as a social evil in order that they could sell the remedy. He identified psychoanalytical differences between the marketing of the soap powders and detergents he considered in the article "Sapanoids and Detergents." Some, like Persil, he thought, based their marketing on the evidence of a *result*, with vanity and social prestige being appealed to by the presentation of a piece of laundry which is whiter than the other, and therefore superior. While others, like Omo, emphasized the *process* of cleaning, and attempted to engage the consumer in what Barthes described as "a sort of experiential mode of substance," implicating the consumer as "the accomplice of deliverance."[165]

In an article titled "Three Kinds of Dirt," Judith Williamson deconstructed the *Hoover Book of Home Management*.[166] She described the three kinds of carpet dirt identified by Hoover and the "particular dangers" posed by each type, then the three cleaning principles that could be used to banish them. She was disheartened at how "the product is wheeled on as the 'answer' to a 'problem,'" while in fact the product itself defines the problem it claims to solve." She wrote: "Each attachment of your Hoover corresponds to some natural function dictated by the very nature of dirt itself!"[167] She drew a parallel between Hoover's marketing practices and those of washing powders that introduced the problem of a "biological stain" in order to provide the solution of a biological washing powder which is required to combat it: "No matter that the washing powder is in fact a *chemical* substance, it must be named to match the stain. The product must be distinguished from its rivals. And it does this by defining the world around it, creating new categories out of previously undifferentiated areas of experience. ... It takes the law to define 'crime'; it takes medicine to define 'sickness'' it takes science to define 'nature'; and it takes Hoover to define the *three kinds of dirt*."[168]

In 1980s design discourse, the notion of taste, presented as an ineffable quality which could be understood only by an elite few, was proffered, like Hoover products or detergents, as a panacea for the perceived lack of taste on the part of the many—a problem most

3.14
In the final section of the "Taste" exhibition at the
Boilerhouse, gray plinths were used (rather than
white plinths or dustbins) to encourage visitors
to make up their own minds about the value of the
exhibits on display. The exhibits in this section included a Russell Hobbs 5452 toaster with honesty
pattern, and James Dyson's Cyclon 1000A pink
vacuum cleaner. Courtesy of The Design Council/
Manchester Metropolitan University.

people did not know they had until it was labeled as such. Bayley's views on the cleansing potential of taste, evident in the exhibition, were even more direct in the press. He was called on with frequency by the Sunday supplements to offer his opinions on what was "in" or "out," "good" or "bad." The same labeling, binary mindset was used by Bayley, York, and Sudjic in their taxonomical or field-guide-like books and essays to help their readers navigate designer lifestyle territory. And it was materialized in the dustbin/plinth device in the "Taste" show. Mary Blume of the *Herald Tribune* wrote: "Both [taste and manners] have been absorbed into the ever changing and repellent notion of lifestyle, and the main thing about lifestyle is that a new set of self-named judges is constantly determining what is good and bad in terms of what is in and out."[169] Bayley told *Sunday Express* readers, addressing them in an exaggeratedly pedagogical and direct second person:

> Every time you buy something you exercise your taste. ... If you think about it you will find that you prefer neatness and restraint. In the end these qualities are more rewarding than confusion and excess. ... Why do you have a gold wristwatch? This metal is inappropriate for the intended purpose. Steel or plastic is better. Perhaps you want to look like a Libyan arms dealer. ... Your choice of the Honesty pattern toaster declares you to be the sort of person who will cheerfully admit, "I love buying cynical junk. Anything the marketing department does is good enough for me." If "country kitchen" is what you want, you'd be better off buying a griddle.[170]

The most emphatic demonstration of the forces of "neatness and restraint" was to be found in Bayley's office, which adjoined The Boilerhouse gallery space and was often considered to be one of the exhibits. The office was designed by Oliver Gregory, one of Conran Associates' founding members, and can be seen as a tangible manifestation of the studied way in which Bayley presented his public image as a modernist and an academic aesthete. He told Fiona McCarthy that he "drank" Pevsner's *Pioneers of Modern Design*.[171] Bayley recalled that in his role as director of The Boilerhouse, he was "part of a missionary campaign to clarify, modernize, and make the world more comfortable, polite and delightful through the application of a chaste version of modern design."[172]

Architecture critic Gavin Stamp, writing in the *Spectator*, described Bayley's office with obvious disdain both for its occupant's exhibitionism and for its high-tech and modernist appurtenances: "The venetian blind in the large internal window of the director's office is always left ever-so-slightly open, so the public can see a carefully posed, High Tec, Clockwork Orange interior. The office is lit by one of those thin, contrived Italian light fittings."[173] Roy Strong, director of the V&A, warned Bayley: "Remember you are not an exhibit even though your office is a lit-up showcase in which you sit, Tussaud's-like, but a human being with passions and feelings and foibles, whose expression explodes in clutter, the true mirrors of humanity and sentiment."[174] Peter York was also fascinated by Bayley's office, and in particular by the curator-on-display phenomenon. He ended his contribution to a BBC Radio 4 piece: "If you go into the corner, there's a special glass box, with an art

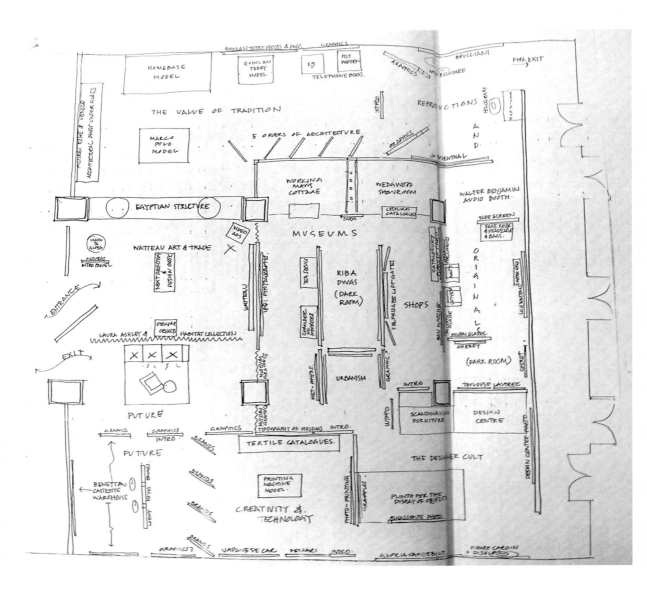

3.15
Stephen Bayley's sketch plan for "Commerce and Culture," the inaugural exhibition at the Design Museum in 1989. Courtesy of The Design Museum London.

DESIGNER CELEBRITIES AND "MONSTROUS, BRINDLED, HYBRID" CONSUMERS

person working, simultaneously reading a magazine and talking on the telephone ... that's the one pièce de résistance, and it's marked 'Young Master Stephen Bayley,' who runs the thing. And that's the real art show."[175]

A full-page article in the *Times*, titled "The Great Taste Test," analyzed the interiors of Bayley's office and home, as well as his personal style. After the guided tour of Bayley's Vauxhall home, which he shared with his wife Flo Bayley, senior graphic designer at Conran Associates, the reporter concluded that Bayley "must have been aching ... for someone to come and write about his taste."[176] If, as Bayley averred, "the major mechanism for establishing good taste is through a small elite of influential individuals who spark off the public's tastebuds," then he was clearly comfortable with being portrayed as an integral part of the "Taste" exhibit.[177]

"Upwardly mobile debris"[178]

By the late 1980s the museum exhibition format, and his own static role within it, began to look rather constricting to Bayley. He became increasingly enamored of other vehicles for expression, especially television. If his Boilerhouse exhibitions were more like magazine articles than exhibits, then the ones he had begun to plan at the Design Museum, which opened in 1989, leaned more toward television as a model. Bayley's plans for the new museum included sharing research costs for the temporary exhibitions with television companies, who would then go to make programs on the same themes. And a Design Museum brochure explained that "the architectural design ... will create a space more like a television studio than a museum: moveable screens, stages, and lights offer a flexible structure."[179] In 1989 Bayley explained: "the process of organizing an exhibition is much the same as making a TV programme, you get an idea, write a script, do a storyboard and interview people. ... The Design Museum will turn its exhibitions into television, creating a far wider constituency."[180]

Indeed, "Commerce and Culture," the Design Museum's inaugural exhibition in 1989, may well have given the visitor the sensation of passing through the sets of a television studio. The exhibition included full-scale reconstructions of the entrance to an American shopping mall, a Corinthian-style column from the Earls Court Sainsbury Homebase store, and Brucciani's gallery of casts from the V&A. There are also parallels to be drawn between this exhibition format and that of some of London's salvage yards, depicted by historian Patrick Wright in his mid-1980s investigations into the tinkers, salvagers, charity shops, and stallholders whose enterprises depended on a close and discriminatory relationship with rubbish. Wright gives us an illuminating portrait of one East End junk merchant's own "jumble of objects," and his *laissez-faire* approach to their curation. Inside his "kaleidoscopic compound," Wright tells us, is a "hastily improvised shanty of temporary structures, each one filled with different categories of upwardly mobile debris." He continues:

The proprietor knows how to pull a likely proposition from its original setting, but he's no master when it comes to reassembly. So his collection mounts up like a great pile of quotations that won't resolve into a single statement, and customers are left to make whatever connections they can. He's got an old, red telephone box, an Ardizzone-style petrol pump, pub signs from some distant hostelry called the Chigwell Arms, a couple of red ticket kiosks for anyone who wants to open their home to the public, a huge chimney pot from a house on Piccadilly, garden seats, wagon wheels that still have the tinsel from some forgotten Christmas hanging from them, sash windows, wrought iron, a couple of fake Doric columns retrieved from a redundant theatre or restaurant design and, of course, lots of stripped pine.[181]

While reviewers of the "Commerce and Culture" exhibition noted visitors' confusion on being greeted by Bayley's eclectic "jumble of objects," and Sudjic, reviewing the show for *Blueprint*, thought it "an anxiety-inducing experience, in which the visitor is assaulted on all sides by music and layered images," that was largely the point.[182] These reconstructions of reconstructions, which were used to illustrate a historicizing impulse evident in postmodern design and architecture, also drew attention, in an embodied way, to the artifice of exhibition-making. And Sudjic did concede that the chief purpose of the exhibition "is to explore just what the terms of discussion of design can be."[183] Before "Commerce and Culture," it was *Blueprint* and The Boilerhouse, of course, which had identified and opened up those "terms," in a vibrant, polemic design discussion, to an expanded audience. Despite self-awareness on the part of *Blueprint*, and an apolitical stance on the part of The Boilerhouse, however, each was a quintessential product of the entrepreneurial individualism espoused by the Thatcher government. Centrally positioned in the nation's capital, within design and architectural practice, and deeply entangled with corporate concerns, neither of these media and museological entities was either inclined or constitutionally suited to provide critical commentary on the social ramifications of consumer practice. Other figures, such as the critics Dick Hebdige and Judith Williamson, more marginally located in design discourse, were more interested and able to attain critical distance on the phenomenon of the designer lifestyle. Their work will be explored in the next section of this chapter.

PART THREE: DICK HEBDIGE'S PATHOLOGICAL AND JUDITH WILLIAMSON'S POLITICAL CRITIQUES OF STYLE

Dick Hebdige's postmodern condition cruise

During the mid-1980s Dick Hebdige lived in Dalston, in London's East End. On the basis of the interest aroused by his 1979 book *Subculture: The Meaning of Style*, in which he had explored the ways in which subcultures appropriated and reconfigured the meanings of images and objects, Hebdige was asked to write for academic journals like *Block* and *Ten.8*, as well as for design magazines like *Blueprint* and socialist publications like *Marxism Today*. He also taught in the Communications Department at Goldsmiths College, where he enjoyed "being in the shadow of practice." He said: "I was always trying to get away from theory, being defined as a theorist." He wanted to write academically and critically, "to be a public intellectual," and because the publications did not pay well, or at all, he used teaching to fund his writing. Additional income came from public speaking, which he was increasingly asked to do during this period. Rather than delivering succinct papers, Hebdige chose a looser, more experimental and performance-based mode of delivery, which he likens to DJ-ing, where he would "stitch" ideas together as a way of "working through, rather than about, something."[184]

3.16
Photograph of Dick Hebdige with Mike Horseman in London, circa 1982, by Paul Edmond. Horseman was a DJ/impresario who promoted punk, Two Tone and New Romantic bands and events at city center clubs such as Barbarella's, the Holy City Zoo, and the Rum Runner, with whom Hebdige worked closely in Birmingham throughout the 1970s, helping him to operate the Shoop sound system on Hill Street in Birmingham. Courtesy of Paul Edmond.

Having grown up in Fulham, London, Hebdige saw himself as urban in orientation, and when it came to choosing a university, he eschewed Oxbridge and picked Birmingham instead, in the second-largest city in Britain. He read English Literature and spent his third year in the Centre for Contemporary Cultural Studies (CCCS) while Stuart Hall was director. Hebdige recalled that Hall took him on as a student on the basis of the ethnographic work Hebdige had begun on pubs in Fulham. In one of these pubs Hebdige had met a charismatic man who was "an artist but also a kind of villain" from a gypsy background who became the subject of his undergraduate dissertation, published in the CCCS Occasional Papers as "Subcultural Conflict and Criminal Performance in Fulham (West London)." His dissertation explored the deployment of the "wind-up"—a linguistic narrative strategy that Hebdige described as "When you're not sure whether what someone says is true or not." He was drawn to what he identified as the use of "coded language," and "silent signs" performed between the man and other pub-goers. He recalled: "This is where my definition of criticality comes from—having this very unstable distinction between play and not play."[185] The game of identifying the coded references of various subcultures was one that preoccupied Hebdige throughout his career.

In 1984 Hebdige suffered a psychotic episode while trying to write an essay on masculinity for the in-house journal of the Labour Party, *New Socialist*: "I was writing this thing and I just got stuck. It was a bit like *The Shining*. I didn't sleep for days. I did automatic writing. And then I jumped out of a window, first floor, and ran off shouting. I thought I was John the Baptist. And the police found me in a giant plastic shoe. It was behind the college, where the carnival stored all their stuff. It was inside this giant boot. And it's got this cross on it, with light bulbs and I thought I was on the cross."[186]

Hebdige was committed to a psychiatric ward, and upon his release he reflected on his breakdown in an essay published the next year in *Ten.8*, "Some Sons and Their Fathers."[187] In the piece he attempted to come to terms with the way in which he had built a masculine identity from fragments of other masculine identities and father figures portrayed in the news, in fiction, in his own life, and in recent cultural memory. Feeling as if, with the breakdown, a narcissistic mirror had shattered, he considered both the example of female role models and the reality of his own father as a way forward in his identity-rebuilding process. The article was a montage of autobiographical, observed, and imagined scenes told in voices that shifted from the public to the personal, and from polemic to narrative accounts of current events such as the miners' strike, Youth Training Schemes, and the deaths of Diana Dors and Alan Lake. As he explained his method: "By trying to speak in more than one dimension—by using different voices and images—I am trying to explore certain possibilities which a more straight-forward approach would, I think, ignore."[188]

Hebdige considered that his breakdown and hospitalization both "broke" and profoundly "changed" him as a writer. He felt that what had happened was "a gift" to his writing: "Maybe it's a romantic thing, not to murder the madman, but to let it come out in the writing." He was interested in developing a mode of writing in which he could channel his own mental instability to achieve a new quality of insight and expression: "I was trying to go in there and do it differently, and come out in a different way. Like you go into the underworld. And to me that's what writing is—you enter into this other dimension. And it is always a risk and an adventure."[189]

In this section we consider two of Hebdige's articles from the mid-late 1980s, paying attention to the ways in which they provided a critique of the ways in which notions of style, lifestyle, and taste were presented to the public via *Blueprint* magazine and The Boilerhouse Project, as well as being illustrative of Hebdige's project in experimental writing.

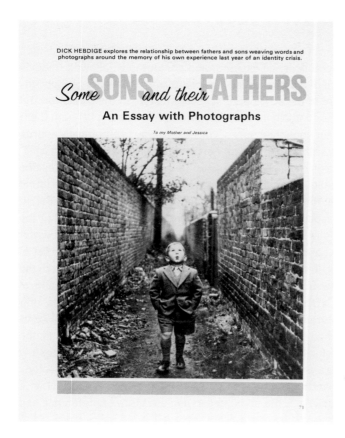

3.17
"Some Sons and Their Fathers,"
by Dick Hebdige, published
in *Ten.8* in 1985. Courtesy of
Dick Hebdige.

Psychosis, schizophrenia, style

In his 1986 essay "A Report on the Western Front: Postmodernism and the 'Politics' of Style," written for *Block*, the left-wing arts journal published by scholars at Middlesex Polytechnic, Hebdige continued to explore the connection between psychosis and schizophrenia and the contemporary preoccupation with style, basing his thinking on the work of theorists such as Jean Baudrillard, Jacques Lacan, and Fredric Jameson.[190]

Jameson, in an argument influenced by Lacan, had drawn a comparison between the postmodern condition and schizophrenia.[191] As Hebdige summarized it: "For Jameson there is the schizophrenic consumer disintegrating into a succession of inassimilable instants, condemned through the ubiquity and instantaneousness of commodified images and instants to live forever in *chronos* (this then this then this) without having access to the (centering) sanctuary of *kairos* (cyclical, mythical, meaningful time)."[192] Baudrillard had also considered schizophrenia to be symptomatic of the postmodern age, and averred that it was not only confusing, but terrifying: "We are now in a new form of schizophrenia. No more hysteria, no more projective paranoia, but this state of terror proper to the schizophrenic. ... The schizophrenic can no longer produce the limits of its own being. ... He is only a pure screen."[193]

The subject matter in "Report on the Western Front," which includes science fiction, urban lifestyles, consumption practices, advertising, and photography, allowed Hebdige to discuss the ideological nature of representation, the confusion of reality or authenticity with hyperreality, and the nature of the schizophrenic "screen" state of the consumer. Yet his aim was not to decode these confusions, to reveal some true meaning beneath them, but rather to glance off and reflect upon their very surfaces as a way to empirically approximate, or to channel-surf, his way through the experience of living in a postmodern age.

The symptoms of psychosis include disorganized thought and speech, delusions, mania, a loss of touch with reality, and hallucinations. Many of these symptoms found their equivalents in the physical form of Hebdige's article "Report on the Western Front." The psychotic state Hebdige associated with 1980s consumerism was embodied in the very structure and texture of his writing. His method was to immerse himself as a writer into the subject matter and to create an authorial character, a particular voice or set of voices, to deal with the material. In the case of this article, he wrote the foreword from the point of view of Ubik, a character from a Philip K. Dick science-fiction novel. Ubik talks of "Dick" (Philip K. Dick) and "dick" (Dick Hebdige) and the way in which the latter was influenced by the novelist's 1978 lecture/essay "How to Build a Universe That Doesn't Fall Apart Two Days Later," in which Dick had discussed his lifelong fascination with Disneyland, the nature of reality, and the authentic human being. The Ubik foreword set up a conceptual frame of reference for Hebdige's article—essential philosophical questions of theology, simulation and inauthenticity—and highlighted the correspondences between it and Dick's essay ("the same limited obsessions ... the same underlying structure of preference and

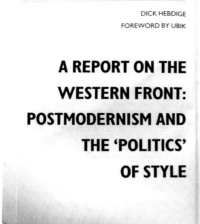

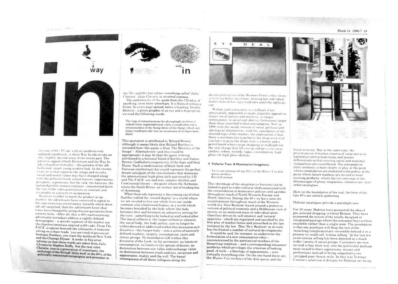

3.18
Spread from "A Report on the Western Front:
Postmodernism and the 'Politics' of Style,"
by Dick Hebdige, published in *Block* 12 in 1986/87.
Courtesy of Middlesex University.

3.19
Spread from "A Report on the Western Front:
Postmodernism and the 'Politics' of Style,"
by Dick Hebdige, published in *Block* 12 in 1986/87.
Courtesy of Middlesex University.

aversion, the same general drift—the scary, funny ride through 'Disneyland' and then the journey home"). In its simulation of schizophrenia through the use of multiple voices, the foreword also established a mood of confused identity, which was a thematic concern in the rest of the article.[194]

After the foreword, the article switches to the first person. This is the voice of Hebdige as academic, speaking both to his audience of art and design students—the article was based on a Bill Chaitken Memorial Lecture he gave at Central School of Art in London in 1985—and to the *Block* readership of his academic peers. It launches with a 270-word sentence, an intentionally unwieldy catalog of the elements of the postmodern "predicament," which extend from "the layout of a page in a fashion magazine" and "the décor of a room" to the "collective chagrin and morbid projections of a Post War generation of baby boomers confronting disillusioned middle age." The literary theorist Ihab Hassan noted how "a new term opens ... a space in language," and indeed Hebdige used the extensiveness of the list to demonstrate postmodernism's status as a contemporary catchall "buzzword," but also its "semantic complexity," its own schizophrenic state.[195]

In order to write about such a multifaceted entity as postmodernism, Hebdige proposed to approach it from an oblique angle, which he said necessitated the article's "eccentric trajectory." The article juxtaposed images, arguments, and parables in an attempt to "reproduce on paper the flow and grain of television discourse switching back and forth between different channels." Much of the work of his critique, then, was done not in the conventional form of a linearly developed argument, but rather through the configuration of the article itself—a distracted assemblage of textual and visual fragments. The sudden scene switches and new topics return the reader to square one at each new section, but as the scenes accumulate they create both an impressionistic portrait of the postmodern condition and a composite argument patchworked from Hebdige's disjointed critiques.

A key linguistic tactic of poststructuralist theorists was wordplay, whereby even the central line of argument could be based on a pun.[196] This was, in part, the legacy of Jacques Derrida's work on language. *Of Grammatology*, translated into English in 1976, was an influential text for Hebdige and his fellow students at the CCCS. Similarly, Baudrillard rarely passed up an opportunity to use punning, assonance, and other linguistic tinkering to draw attention to the flexibility and multiple meanings of language, as well as to the surface of his text. Hebdige channeled some of these tendencies, especially when he wrote about Baudrillard: "In the (ob)scenario sketched out by Jean Baudrillard ... the metaphor of television as the nether-eye (never I.)" Hebdige even commented on himself doing it: "Somewhere in the middle, between the seminar and the cinema sits the work of Jean Baudrillard (the rhyme seminar/cinema/Baudrillard is an irritating if apposite coincidence ...)."[197] His self-reflective incursions interrupt the flow of the article, forcing the reader to share in the experience of working through ideas with the writer.

While the aim of "Report on the Western Front" was to "cruise the postmodern condition" in its entirety, within this broader purpose Hebdige focused specifically on the ways in which style and lifestyle epitomized aspects of postmodernism, in ways that connect and contrast to the other writers discussed in this chapter. "There are plenty of signs of the Post on the frantic surfaces of style and 'lifestyle' in the mid to late 80's," he wrote, as a way to narrow his field and to introduce another more positive view of postmodernism which connects to the concerns of criticism: "it often gets depicted ... as a celebration of what is there and what might be possible."[198] Hebdige wrote of "a growing public familiarity with formal and representational codes, a profusion of consumption 'lifestyles,' cultures, subcultures; a generalized sensitivity to style (as language, as option, as game) and to difference—ethnic, gender, regional and local difference: what Fredric Jameson has called 'heterogeneity without norms.'"[199]

Hebdige referred to an "Ideal Consumer" of the late 1980s as an "it," stripped of personal pronouns, in reference to the latest urban fashion for transgender experimentation. He described this ideal consumer as "a bundle of contradictions: monstrous, brindled, hybrid":[200]

> a young but powerful (ie. Solvent) Porsche owning gender bender who wears Katherine Hamnett skirts and Gucci loafers, watches Dallas on air and Eastenders on video, drinks lager, white wine or Grolsch and Cointreau, uses tampons, smokes St Bruno pipe tobacco, and uses Glintz hair colour, cooks nouvelle cuisine and eats out at McDonalds, is an international jetsetter who holidays in the Caribbean and lives in a mock-Georgian mansion in Milton Keynes with an MFI self-assembled kitchen unit, an Amstrad computer and a custom-built jacuzzi.[201]

Hebdige's characterization of an impossible being, indulging all of its contradictory desires, as well as its national, cultural, class, and sexual identities with a motley of conflicting brands and lifestyle choices, points to a schizophrenic state, where fantasy and reality collide in a dystopian orgy of consumer choice, symbolic of the postmodern condition: "It [the postmodern consumer] is a complete social and psychological mess."[202] This account of dual-gender consumer values provides a marked counterpoint to the narratives of consumption presented by The Boilerhouse and *Blueprint* which, when they did consider the use of designed products, nearly always privileged a male viewpoint.

In a section of the article that deals with what Hebdige terms "A Monetarist Imagery," he analyzed the Habitat catalog, which, first introduced in 1966, was one of the furnishing company's primary marketing tools. In 1983, design writer James Woudhuysen wrote about the Habitat catalog and Terence Conran's role in nurturing a consumer base in Britain for clean, modern design. Woudhuysen described the catalog as being "thick with pastelshaded blinds, jolly Anglepoise lamps and tables that look so wholesome and chunky they could almost double as chopping boards." In explaining the catalog's role in facilitating Conran's mission to improve the taste of his potential market, Woudhuysen wrote: "The Brixtonians buy it; so, every year, do a million other people in Britain. It has been designer and entrepreneur Sir Terence Conran's singular achievement to find them and train them to trust his sense of form, line and colour, come what may."[203] Where Woudhuysen's account suggests only skepticism of Conran's role as a "trainer" of the public, Hebdige's reading of the catalog and Conran's influence was much darker, and illustrates how his perspectives clashed with those of most design journalists of the period. Hebdige saw the catalog as a paradigmatic example of a "consumer aesthetic which privileges the criterion of looking good, of style—a theology of appearances—over virtually everything else."[204] He considered the Habitat catalog, like glossy magazines, commercials, and mail order catalogs of the period, as constituting a "dreamscape" in which "future markets are invited to meet existing products." (This concept recalls Richard Hamilton's evocation in "The Persuading Image" of manufacturers in 1950s America "moulding" consumers to fit products they had already created, as discussed in chapter 1.) Hebdige credited Habitat with pioneering what he called "syntax selling"—where consumers were encouraged to buy into a particular lifestyle by purchasing a complete ensemble of furniture and products.[205]

Hebdige compared Conran's ability to provide niche products for emerging niche markets to Pierre Bourdieu's concept of "'habitus'—the internalized system of socially structured, class-specific gestures, tastes, aspirations, dispositions which can dictate everything from an individual's 'body hexis' to her/his education performance, speech, dress, and perception of life opportunities."[206] Acknowledging that Conran's goal was to generate profits, to educate the public, and to raise the general standard of design in Britain, Hebdige also observed: "it may also incidentally lead to the development of the 'cultivated habitus,' a 'semi-learned grammar' of good taste which would serve to perpetuate a hierarchy of taste by establishing a scale ranging from excellence (mastery of the code), the rule converted into a habitus capable of playing with the rule of the game, through the strict conformity of those condemned merely to execute, to the dispossession of the layman."[207]

The room settings and complementary ensembles of household items on display in the Habitat catalog provided the consumer with the "security and imaginary coherence of pre-scripted life style sequences," Hebdige asserted. This type of marketing was, in his view, a form of "institutionalized therapy for the psychotic consumer," which he imagined thus: "This is the chair to sit in, the food to eat, the plates to eat it off, the table settings to place it in, the cutlery to eat it with. This is the wine to drink with it. These are the glasses to drink the wine in, the clothes to wear, the books to decorate the bookshelves with. Now that Conran has taken over Mothercare, you can colour co-ordinate your entire life from cradle to grave."[208]

The soothing rhythm of this passage, with its repeated clause "This is ...," in the voice of someone speaking to a mentally ill patient or a young child, cast the lifestyle shopping experience as a form of therapy for the very condition to which it gave rise. Hebdige was particularly concerned about the "lack of local resistance" to these increasingly sophisticated marketing strategies due to the "spread and penetration of market values," enabled in part by *Blueprint* and The Boilerhouse.[209]

Reflecting on the role of criticism, however, Hebdige questioned whether his own criticism was always about resistance. "It's also about articulation, about creating bridges, and orchestrating transitions, imagining another way of moving forward," he said. "You're actually giving a prescription, which is also like a piece of marketing, really. ... You have a role to play in shaping opinion ... it's not about saying 'no' all the time."[210]

"Shopping-Spree in Conran Hell"

In "Shopping-Spree in Conran Hell," first published in *Block* and then republished in an edited version as "Shopping for Souvenirs in the Occupied Zone" in *Blueprint* in 1989, Hebdige recounted a 1988 trip to Eastern Europe through his observations of shop windows and consumer behaviors, and his own "captured images" and souvenirs.[211] He contrasted the lackluster experience of consuming, or attempting to consume, in the Eastern Bloc just prior to the fall of the Berlin Wall, to the excesses of Western shopping habits and in particular to those typified by Habitat stores, likening Poland to a "Conran Hell," where objects "look and feel as if they've fallen into the material world from some more shadowy dimension."[212] Hebdige's own "hunt for souvenirs" acquired an "unsavoury patina," he wrote, when it became clear "that most local people have to spend a large part of their waking lives hunting down the bare necessities, the most minimal kinds of luxury goods."[213] He returned from the trip with:

> a spring hipped cardboard suitcase bought in Prague filled with literal souvenirs—a golden plastic saxophone made in Russia, a heavy Czech military issue combination cork screw/can opener in no-nonsense steel, a genuine zinc samovar, a rare half-melted tablet of soft greasy Polish hotel soap, a plastic spoon the colour of fresh egg yolk from Czechoslovakia Air Lines, an assortment of documents: visa and currency exchange stubs, hotel bills, museum, cinema and tram tickets.[214]

Hebdige's list provides a critical counterpoint both to Banham's catalog of exotic American "goodies" in his 1963 autobiographical article "Who Is This Pop?" (discussed in chapter 1), and to the numerous lists of expensive designer objects deemed essential to the construction of a designer lifestyle in 1980s London, and displayed on plinths at The Boilerhouse. By importing these mundane Eastern bloc objects and ephemera into the pages of *Blueprint*, Hebdige confronted the magazine's readers with the realities of privation beyond their Western capsule of privilege, and offered a politicized riposte to the fetishization of luxury goods, which was the regular fare of the magazine. And yet, through his addition of luxury-conferring adjectives such as "golden," "genuine," and "rare," Hebdige ended up romanticizing the objects, and almost undercutting his political intention.

Hebdige observed the relationship between "goods and cultural values" in the dressings of shop windows in which he saw a soured shadow of the American dream of consumption. In Warsaw, where "scarcity makes for a more generally desolate dreamscape pierced by the odd transcendental shaft of purist aspiration," he noted:

Window shopping here takes on an ethereal quality which is enhanced by portraits of the Pope which smile benignly down on empty spaces, dusty glass from the walls of the shop interiors. The typical display: a few items—some hats, or shoes, a doll, a box of unidentified machine parts—are placed against a faded curtain or a piece of paper complete with drawing pins and yellowed in the sun. In the window of one clothes shop they've given up pretending that looking and buying are in any way related. The window is empty except for an old copy of *Vogue* from the late 70's. It lies open in the centre of the window: a sign of a dream or a dream of consumption which may have taken place some time ago and somewhere else.[215]

Hebdige's evocation of the entropic character of consumption was not confined to the Eastern Bloc; in the West, too, in his view, the satisfaction supposed to follow from buying things was inaccessible to most. Similarly, Czech interior designer Eva Jiricna said, of coming to London in 1968, "When you first arrive, you are absolutely amazed by being able to choose. You can select any one of 200 carpets, or thousands of bricks. It takes you years to realize that most of them are junk."[216] In "Western Front," Hebdige described the claustrophobic nature of a consumption-driven society in which shops represent both the source of discontent and the only available public space for expression of that discontent: "Now in 1986 with the steady erosion of social, political, and ideological alternatives, with the ascendancy of the stunted logic of the market, the implication is that there is nowhere else to go but the shops even if all you have to go to the shops with is a bottle and a petrol bomb when you go shopping at midnight for the only things that lift you up and give you value: clothes, videos, records, tapes, consumption: high gloss i-d, high gloss identity."[217]

Judith Williamson had a less fatalistic view of shopping, at least of the social potential of shopping. Even though she believed that products are used by consumer society to "channel" and "contain" extreme emotions such as passion, she admitted that "consuming products does give a thrill, a sense of both belonging and being different."[218] She wrote: "Consumerism is often represented as a supremely individualistic act—yet it is also very social: shopping is a socially endorsed event, a form of social cement. It makes you feel normal. Most people find it cheers them up—even window-shopping."[219] And in her introduction to *Consuming Passions*, she conjured "Christmas trips of childhood to Oxford Street" where, in the lighted windows, she had seen "passions leaping through the plate glass, filling the forms of a hundred products, tracing the shapes of a hundred hopes."[220]

Judith Williamson: redirecting emotions from objects to actions

Although she addressed the same kinds of subject matter that Dick Hebdige did, the socialist cultural critic Judith Williamson approached it from a more stringently defined political and class-conscious angle. Her feminism and Marxism were both explicit and implicit in most of what she wrote, and, in line with her politics, she sought a broader audience for her writing, choosing wide-circulation publications such as *Time Out* and *City*

Limits over academic publications such as *Block*. Like Hebdige's, Williamson's critiques of consumer culture targeted Thatcherite values, but she was equally critical of the political Left, as represented by the British Communist Party (which she saw as having co-opted style as a means of rebranding) and the left-leaning academic community (which she saw as having embraced cultural studies, and in particular, style, grateful for the "softer" territory of the superstructure and in an attempt to align with contemporary fashion).

In the 1980s Judith Williamson lived on a council estate in Tufnell Park, North London, and was closely involved in community politics. Later, she would be recognized with a Mayor's Civic Award for her work for the Brecknock Road Estate Tenants & Residents Association, which she founded in 1983. Her parents were from different class backgrounds. "My father was working class and my mother was from a very upper-middle-class background," a disparity that she thought gave rise to her "political sense of aesthetics."[221] Williamson studied English Literature at Sussex University in the School of English and American Studies, with a final year at the University of California, Berkeley. The work she did at Berkeley, developing a semiotic analysis of advertisements using clippings she had been collecting since she was a teenager, was published in 1978, when she was only twenty-two, in the book *Decoding Advertisements*. Williamson would continue this practice of filing clippings from newspapers and magazines. Her shelves still house file boxes with labels such as "Riots 1981," "Royal Wedding 1981," "Falklands 1982," "Madonna," "Production," "Ads—Gender," and "Sellafield/ Nuclear Power," for example, in a tangible record of her interests and methods. She also collected issues of the magazines she wrote for, as well as other titles such as *Viz* and *Spy*, and Habitat catalogs.

In 1982, when she graduated from the Royal College of Art with an MA in Film and Television, Williamson began writing film criticism for *Time Out*. In its early days this weekly London listings magazine was run on cooperative principles, with staff members paid the same amount (£8,500) whether they were receptionists, typesetters, or writers. In 1981, when the management decided to introduce a sliding pay scale, the staff went on strike, creating an ad hoc publication supported by donations from the public. Williamson recalled that she felt as if she were in direct communication with her audience: "I was ... aware of no longer being an anonymous commentator on movies, but being in a situation known to every reader of the broadsheet, and I learned one of the first lessons of journalism—your readers are real."[222] She later added that in finding out she was not writing to herself, "There was a sense of liberation and for me, perhaps a loosening up of style and tone, which lasted through the rest of my time as a critic."[223]

The group failed to win the strike, but set up *City Limits*, a rival listings magazine organized as a cooperative. The launch issue's editorial stated: "Six months, innumerable dismissals, several writs, threats, recriminations, sit-ins, lock-outs and undignified rumbles later, we have brought you *City Limits*—a paper that we think you'll agree was worth the fight."[224] The graphic designer David King created what Williamson considered a "bold, quasi-constructivist design" for the publication that reflected its alternative viewpoint:

"We looked oppositional," she reflected.[225] Williamson continued at *City Limits* as a staff member, also teaching in the History of Art, Design, and Film department at Middlesex Polytechnic—a role which she saw as having contributed to her sense that "explaining is a big part of criticism"—even as she took up a new post as film critic, responsible for a weekly column in the *New Statesman* starting in 1986.[226]

Williamson, more than most critics of the period, was emphatically clear about her political stance. "I came into writing and thinking as a fully formed Marxist with a critique of the way the world is," she said.[227] While Hebdige and Sudjic used ambiguity and multiple voices as writerly modes, for Williamson, declaring one's "position" was fundamental to the practice of criticism.[228] She used an early column in the *New Statesman* to articulate this position so that her readers would know exactly how to interpret her commentary on film, telling them: "It should be clear to anyone who has read this column over the last few months that I am writing with a feminist and a Marxist politics ... a political view of cinema [can] provide ways of questioning assumptions about the structure of society, of challenging what we take for granted."[229]

Williamson was aware of what she saw as a contradictory impulse in criticism between the exercise of personal taste and the idea of absolute values—that "critics' judgments are seen as at once totally personal, and yet—paradoxically—profoundly objective." She wrote: "I have tried to suggest that the 'personal,' supposedly random nature of taste effectively depoliticizes it, takes it away from the realm of class. But the other side of this contradiction, the idea of inherent value, plays a key role in *maintaining* what amount to class divisions in the realm of Culture, where some products are seen as infinitely more 'value-ful' than others."[230]

Williamson's writing negotiated these poles, yet its personal nature is striking. "It is impossible to write regularly, week after week, under intense pressure, without feeling that you are squeezing a little bit of yourself into it all the time," she wrote of her work as a film critic.[231] The self that she squeezed in was manifest in her always-present political filter, but also in anecdotes and images from her daily life. In a *City Limits* piece on the need for socialists to fight for social and public life, instead of personal ownership, she wrote: "the sense of Welfare State is one of the earliest things I can remember, the delicious Clinic Orange Juice that was quite unlike 'bought,' the equally foul Cod-Liver Oil, the reverence with which my father spoke of Nye Bevan, and the idea that the world was supposed to get better."[232] What elevated this kind of personal writing beyond the merely anecdotal or "quirky," for Williamson, was its potential to connect with her audiences and to provide them with a means for performing criticism themselves, to give them "access to intellectual structures" whereby they might "make their *own* critical judgments and decisions."[233] She wrote: "I live and work within the same culture that produces the films I write about; my feelings and reactions may be my own, but they are not necessarily *only* my own."[234] Similarly, when reflecting back on design criticism of the period, she said: "People who write about design aren't fuelled by different drives from anyone else."[235]

As a socialist and a Marxist, Williamson was critical of style—which she saw as a manifestation of capitalist, and particularly Thatcherite, culture—and the way it was idealized in the design press.[236] She was also frustrated by what she saw as the academic Left's soft engagement, and seeming infatuation, with style—its lack of a more "daring socialism."[237] She distanced herself from both camps, preferring instead a fast-paced schedule of weekly columns for widely read publications. She explained that she "was coming from the approach of someone trying to understand design culture. At that time design was hot. The idea of the designer object emerged right then. The idea of a lifestyle was central to the early 1980s idea of consumerism and consuming designed objects that would speak about you."[238] This outsider status, constrained neither by friendships with designers nor by any economic ties to the industry, may explain Williamson's ability to achieve critical distance in her writing about design.

Williamson's writing lacked the fizz and verbal dexterity of the New Journalism-infused style found in *Blueprint* and the citrus snark of the cultural commentator Julie Burchill in *The Face*, but it engaged and persuaded through its forthright conviction and clarity. She maintained her own distinctly nonacademic language, summoning Marx, Freud, Barthes, and Benjamin only when necessary to give ballast to her points. She strongly believed in the power of writing; language for Williamson, in 1984, was "the only power we have left in the undeniable world of consumerism."[239]

Her reference base was drawn from a London-centric urbanscape of communal experiences in buses, housing estates, and public parks. Depicting the joys of spontaneous community experience, in a 1984 essay about the Walkman, she wrote: "There is a kind of freedom about chance encounters, which is why conversations and arguments in buses are so much livelier than those of the wittiest dinner party. Help is easy to come by on urban streets, whether with a burst shopping bag or a road accident."[240] Her readers must have been convinced that when Williamson wrote of the social dynamics of housing estates and public transport, these situations were lived experiences rather than detached, writerly observations. The characters that figured in her articles gathered in one another's living rooms to watch TV programs like "Dallas," went to the cinema, and wore legwarmers. Unlike the characters portrayed in the pages of *Blueprint*, they didn't go to the Milan Furniture Fair or wear Rolex watches. Williamson used humor and emotional persuasion to make her points, but mostly her writing was serious and concerned with what might be done at the level of grassroots activism.

Williamson's writing on design, consumption, and lifestyle represents a more politically motivated take on such subject matter than we have seen in the other writers discussed in this chapter. She demanded more of design than *Blueprint* did, but she also thought that the cultural theorists of the time, whom she saw as preoccupied with meaning at an abstract level, could have used a little more of *Blueprint*'s concreteness in their work.

"Urban Spaceman"[241]

Williamson's essay "Urban Spaceman," written in 1984 specially for her *Consuming Passions* collection, considered the Walkman not as a designed object per se, but rather through the way its image was advertised, the way it was used, the way it shaped or altered public space, and its broader symbolic meaning as a reflection of an increasingly individualized culture of the kind engendered by the Conservative government. She wrote that the Walkman "provides a concrete image of alienation, suggesting an implicit hostility to, and isolation from, the environment in which it is worn. Yet it also embodies the underlying values of precisely the society which produces that alienation—those principles which are the linchpin of Thatcherite Britain: individualism, privatization, and 'choice.'"[242] Williamson was not being paranoid; she was well-attuned to the politics of the day. The historian Robert Hewison recounted of the period that, through Thatcherism, "The British soul was to be remade, by creating a new myth of economic individualism to replace the old ideals of community and collectivism."[243]

Williamson's depiction of the Walkman differs dramatically from those of other design writers. Sudjic heralded the Walkman as a "cult object" along with the Zippo lighter, the Mont Blanc pen, and other status-conferring products.[244] And Bayley—who put the device on display as part of a 1982 exhibition devoted to Sony at The Boilerhouse, and wrote at length about its genesis in the accompanying catalog—returned to the Walkman in 1985, describing it as "perhaps the definitive consumer product of the eighties, another example of Sony's remarkable flair for innovation."[245] A gushing newspaper article, which leaned heavily on the Boilerhouse-issued press release about the Sony exhibition, proclaimed: "Among the extraordinary exhibits at the Boilerhouse was the latest Walkman person radio. The Walkman is one of the happiest inventions of modern times, since it allows music fans better quality than ever without causing others the fury of being forced to listen to music they don't like."[246]

Baudrillard also discussed the Walkman, a product that both fascinated and repelled him. In 1986 he saw its use, in conjunction with the fitness fad of jogging, in apocalyptic terms: "Nothing evokes the end of the world more than a man running straight ahead on a beach, swathed in the sounds of his Walkman, cocooned in the solitary sacrifice of his energy."[247]

For her part, Williamson discussed the Walkman in terms of the ways it reshaped urban space, anticipating a focus on design's social and political context that design criticism would go on to engage with in the coming decades: "I was profoundly interested in [designed products] as physical objects which organize space and organize behaviour. The way you use an implement is going to be partly determined by its design and shape and with public spaces the ways they are designed and organized make people move or sit in particular ways."[248] In "Urban Spaceman," she depicted the Walkman as "primarily a way of escaping from a shared experience or environment. It produces a privatized sound, in the public domain; a weapon of the individual against the communal."[249] And unlike those who preferred the Walkman to the use of "squawking suitcases," Williamson recommended the ghetto blaster, which she thought helped create "a shared experience, a communal event."[250]

The Politics of Consumption

While most of Williamson's articles ended up being about the politics of consumption, one, published in *New Socialist* in 1985, addressed the topic head on. The article drew attention to the way in which the needs and desires that fuel consumption were "both sharpened and denied by the economic system that makes them."[251] The article was also about how she thought left-wing writers should write about products—in particular, why they should identify the ideologies and economic realities that drive their consumption: "The analysis of consumer items as the concrete forms taken by particular needs is essential if socialists are to envisage different ways of meeting them."[252]

Her critique was directed in part at recent writing about the "lifestyle craze" she saw in publications such as *Blueprint*. Journalistic writing about consumer goods was implicitly to blame because it dealt only with their forms, which are, she said, "fundamentally those of market capitalism," and did not deal with the "needs that underlie use."[253]

In an essay titled "Belonging to Us," written for *City Limits* in 1983, Williamson used a section of an election broadcast by Margaret Thatcher about the Conservative value of property ownership to discuss a social situation in which possessions and the concept of home had become "more than ever a symbol of yearned-for security."[254] The tragedy was, she wrote, "that as this right-wing government makes ordinary life harder and harder, it creates the social conditions for precisely the individual fears and anxieties which fuel its support."[255] Home, to Williamson, was not the staged room settings of a Habitat catalog, nor the minimalist interiors featured in *Blueprint*. Home was constituted not by belongings, in the sense of things owned, as she thought Tory individualism would have British society believe, but rather by the feeling of belonging to a place—in her case a shared, public, urban place like London.

If her critique of Thatcherism and design culture's collusion with its values was uncompromising, she was equally tough on the Left: on the cultural theorists' over-eager embrace of postmodernism, and the Communist Party of Great Britain's seemingly uncritical adoption of the style phenomenon for its own rebranding purposes: "What ought to have been opposition in many parts of the Left, what should have been a left wing politics offering something different, actually went with the flow and moved into the lifestyle mindset."[256] For her, one incident in particular encapsulated this unsettling tendency. On Remembrance Day in 1981, the Labour Party leader Michael Foot wore a black duffel coat to the annual wreath-laying ceremony at the Whitehall Cenotaph. Williamson recalled: "There was an outcry. He was supposed to have worn a smarter, more stylish coat. The media pounced on him. The left-wing writers who should have been there putting the argument that style doesn't count in this context, didn't. The coat became symbolic of what this new trendy, cultural studies-influenced individualist left interested in identity politics wanted to cast off. Post-punk stylists were saying that the left should smarten up."[257]

Williamson was impatient with the "post-punk stylists," the champions of cultural studies—Hebdige included—and members of the 1960s Left who had recently "discovered" style and who portrayed consumerism as a "progressive trend" where commodities or styles can be "subverted."[258] She was particularly dubious about postmodern readings of culture, in which meaning was unfixed, and "one can claim as radical almost anything provided it is taken out of its original context."[259] Postmodernism rankled with her because it bred what she saw as the lazy use of theory in academia: "You can apply the same term to a building, a political party, or a hairstyle, without, apparently, the slightest need for modification."[260] It was also, in her view, a conspicuously male-dominated field: "Why is so much of the 'serious' stuff on postmodernism written by men? Especially when pm is supposed to be all about the feminine, the other, dispersal, difference."[261]

To Williamson, Leftist cultural writers were too caught up with consumerism and not interested enough in the failing sphere of British production. In a post-Fordist era, in which society was becoming increasingly disconnected from the means of production, she posited the examples of miners' and printing union strikes (between 1984 and 1986) as important attempts on the part of working-class citizens to regain control of their products, environment, and communal identities, and therefore as important subjects of study. But most writers seemed to her to find street-style struggles more "riveting" than labor struggles, thus the widening gap between production and consumption was left unchallenged.[262]

For this state of affairs Williamson blamed Jean Baudrillard, whose ideas had captured the imagination of cultural theorists, style-conscious youth, and journalists alike. Writing in 1988, she observed: "He has become the prophet of the style era—and with good reason: for his writing perfectly describes the world of, for example, *The Face*, and it is little wonder that the world in turn looks to him as its guru."[263]

Her frustration with Baudrillard lay in his increasing rejection of the possibility of a world beyond the simulacra. Williamson was still guided by Marxist and psychoanalytic thinking, and saw continued relevance in the "depth model" that sees the surface of cultural signifiers and images to be an ideological distortion of operant forces "below," which can be excavated through ideological critique. When Williamson interviewed Baudrillard for *City Limits* in 1988, she asked him to locate the space from which an evaluation might take place of what he had called "the double challenge of the masses and their silence, and of the media." He responded that, according to his conception, "there is no longer any possibility of evaluation. ... There isn't any point of view from which to criticize [the masses] external to that space," confirming her apprehension that the notion of "seduction" had indeed replaced "interpretation."[264]

At the end of the 1980s, *Marxism Today*, the British Communist Party's journal, which had been edited by Martin Jacques since 1977, drafted a manifesto and commissioned a series of articles and responses around a movement they dubbed "New Times." In this special edition, published in October 1988, *Marxism Today* was forced to admit that "increasingly, at the heart of Thatcherism, has been its sense of New Times, of living in a new era. While the Left remains profoundly wedded to the past, to 1945, to the old social democratic order, to the priorities of Keynes and Beveridge, the Right has glimpsed the future and run with it. As a result, it is the Right which now appears modern, radical, innovative and brimming with confidence and ideas about the future."[265]

Williamson was skeptical of such a reading, believing that the Left had got caught up in misguided reverence for the dynamism and populism of the Thatcher government, and abandoned wholesale its own socialist traditions. She wrote a response to the "New Times" manifesto for *New Statesman & Society*, attempting to explain her unease with the document by studying it symptomatically. The important question for Williamson was: "why has the market place been such a powerful platform for both right and left wing rhetoric?" She acknowledged that "the appeal to voters as consumers is a powerful one because it recognises people's needs for pleasure," but pointed out the problem that "shopping for democracy" lay in the unequal distribution of the means to do it, "plus the appalling conditions and pay of the workers in places like South Korea where so many of our lifestyle accoutrements are produced."[266] She took issue with the fact that "New Times," referred to by *Marxism Today* in the singular, implied that it was one inflexible entity, rather than a multitude of views. From this standpoint, those who wanted new ideas for the future, but without losing socialist traditions, were constrained by the rigidity of the program: "Any wish to maintain a link with the past is portrayed as 'hankering.'"[267]

While Williamson, in her critiques of both the Left and the Right's engagement with style, appears to have been caught in a stalemate situation, Hebdige was ultimately more optimistic about the future of criticism and the possibility of articulating "a new kind of socialism." In his own contribution to the "New Times" discourse, he concluded an article about postmodernism's relationship to the newly conceived socialism by averring: "Contrary to what Baudrillard says [decadence is the yearned for end of everything], there is nothing fatal or finished about the new times. The task for the 90s has to be how to rise to the challenge, how to abjure certain kinds of authority we might have laid claim to in the past, without losing sight of the longer-term objectives, how to articulate a new kind of socialism, how to make socialism, as Raymond Williams might have said, without the masses."[268]

CONCLUSION

By the end of the 1980s, the British design boom that design commentary of the period had both fed, and fed from, was imploding. Several of the large design consultancies collapsed. Michael Peters, who had been one of the initial funders of *Blueprint*, experienced a pretax loss of £2.94 million in the six months to December 1990. Fitch, another *Blueprint* funder, saw its share price dropping, and Conran's Storehouse was also in trouble.

In Sudjic's 1993 assessment of design's rise and fall, the greed associated with the design boom had finally consumed itself: "Like Tom and Jerry running over the edge of a cliff, their paws whizzing round like propellers until they finally looked down, smart young developers continued to invest in property and designers continued to go public. The building societies and the banks fell over themselves to fund it all, and the economy was awash with cash and Starck chairs as a result. Then the sky started to fall in."[269]

As the design decade drew to a close, and, as the Memphis collective member Barbara Radice had predicted, the "glitter" began to "fade and liberate, as radioactive fallout, the shivers and omens of the end," a palpable weariness became evident among the editors, critics, and curators considered in this chapter, in the face of what they saw as an uncheckable velocity in the turnover of fashions, and the increasing rate of obsolescence of their own media products—their exhibitions and magazines.[270] They also became more reflective and introspective. Sudjic at *Blueprint* began to consider the ecological impact of design's production processes, to question his role in the design star-making system, and to advocate improved historical knowledge in design criticism. In his 1988 summary of the decade, he asserted: "If design criticism is to have any usefulness at all, it must be to draw attention to this phenomenon [the way in which designers turned out styles], to remind designers of the need for a sense of history. It's been a decade in which design has sought to discover a critical and theoretical underpinning for what it does. After decades of depending on architectural discourse, design has tried to strike out on its own to find a sense of direction. And design theory and history has burgeoned as an academic study."[271] Sudjic was referring here to the emergence of design history as a discipline which gained momentum in parallel to, and partially enabled by, the design decade. Fostered by the Design History Society, founded in 1977, and the V&A/RCA History of Design MA course, founded in 1982, design history connected to the design boom, through such links as its Fitch and Company scholarship, which funded students who aimed to work in the field of design management or industry.

Stephen Bayley, too, in his description of the 1989 "Culture and Commerce" exhibition, the first to be held at the new Design Museum on Butler's Wharf, sounded jaded when he noted: "Metaphorically, 'designer' has become a journalistic cliché." He devoted a section of the exhibition to the subject of the "Designer Cult," a notion which he amplified in strident, if reformist-sounding, terms in the exhibition catalog: "The designer cult, with all its pompous absurdities, travesties of value and its short lived pretensions, is a sort of revenge of tradition of the carefully nurtured culture of mass production."[272] He continued in the same dyspeptic tone: "Hitherto separate, these gauges of consumption (style, fashion, taste) were all rolled in to the 'designer,' one of the silliest (and most transient) manifestations of postmodernism's dedication to expensive trash."[273]

3.20
Front page of the *Guardian Review*, December 2, 1988, created by Jon Wozencroft and Neville Brody. Courtesy of Jon Wozencroft.

"I was overwhelmed with a sense of futility," Bayley recalled of his state of mind in the summer of 1989, when he relinquished his directorship of the Design Museum only a month after its official opening. He remembered thinking: "What have we done? We've spent all this money and we've just recreated the Conran shop. Personally, I had spent eleven to twelve years setting up the Design Museum. By the time it opened, design was no longer this noble world-improving calling with a very clear aesthetic. It had become a synonym for anything meretricious, expensive, odd and curious which was never my intention. Which was part of my falling out with it all."[274]

The writer Jon Wozencroft and designer Neville Brody seemed to have fallen out with design too. In their anti-style manifesto, published on the front page of the Review section of the *Guardian* on December 2, 1988, they gave typographic form to what they perceived as the erosion of design's status: "DESIGN, Design, dsgn. The word itself has grown tiresome."[275] This succinct obituary for the preoccupations of an era also served as a coda for the end of, or a pause in, the forward thrust of design commentary. What had been a "scratch of an exhibition centre" became a fully fledged, government-funded museum in 1989, and what had been an ad hoc and irreverent magazine, with its every detail controlled by its publisher and editor, "lost its way," as Peter Murray put it, and was sold in 1994.[276] *Blueprint*'s spotlighting of design icons and iconic designers, and fascination with the designer lifestyle, all became recognizable tropes dispersed throughout the journalistic domain. The Boilerhouse's compulsion to dictate good taste, and to "Hoover" up or "Dust-Off" bad design, spread well beyond its ceramic bunker. Such innovations and impulses were incorporated and intensified in design media of the 1990s, typified most obviously in a magazine like *Wallpaper*, launched in 1996, with two of Hebidge's "monstrous brindled hybrid" consumers depicted on its cover, now recast as "urban modernists." Hebdige's use of theory and Williamson's use of politics enriched design criticism, and their concern with the social and psychological effects of design suggested multiple routes for its diversification. But the design media of the 1990s, progeny of *Blueprint* and The Boilerhouse in many respects, found they had very little space or time for extended critique of the international jet-set lifestyle they were immersed in. While *Wallpaper* heralded the "global nomad" as the social archetype of the 1990s, design criticism found itself increasingly homeless. A growing sense of the futility of critical judgment—of language, even—engendered a silence in critical discourse in 1990s Britain, out of which emerged two nonverbal alternative modes of design criticism, to be explored in the next chapter—the exhibition, and the designed product itself.

Left:–
'The Boardroom'—Michael Marriott's
studio.

Below:–
Drawing for 'Ruth' trestle table, 1997.

Michael Marriott
4 & 6 Ellsworth Street, London E2 OAX. Tel & Fax *44(0)171 613 5581

Michael Marriott

b.1963
1993 MA Furniture Design
Royal College of Art, London

Michael Marriott designs furniture and
products with no stylistic preconditions,
and with which the user can engage on
an uncomplicated emotional level.
Characterised by an attention to detail,
Marriott's designs are honest and simple.
Furniture is based on generic shapes;
unsophisticated materials such as MDF
and pine are used, and designs
sometimes incorporate or refer to
familiar or found items—a plastic bucket
is a light fitting, wooden spoons become
coat hangers, skittles are table legs. The
scale is compact, the design is utilitarian
and flexible. All elements draw on
notions of the ordinary and everyday and
are used to introduce a spark of
recognition. Occasionally designs appear
endearingly ugly, an invitation to users
to investigate their own ideas of beauty
rather than be dictated to by the fashion
and 'style mafia'.

53

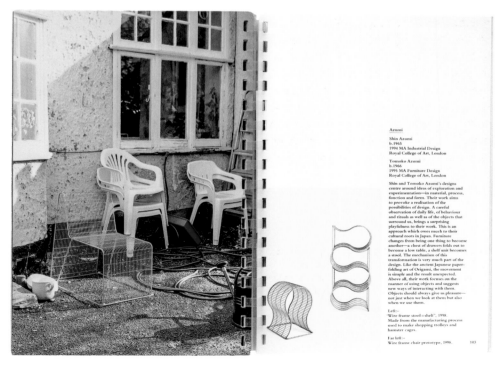

Azumi

Shin Azumi
b.1965
1994 MA Industrial Design
Royal College of Art, London

Tomoko Azumi
b.1966
1995 MA Furniture Design
Royal College of Art, London

Shin and Tomoko Azumi's designs
centre around ideas of exploration and
experimentation—in material, process,
function and form. Their work aims
to provoke a realisation of the
possibilities of design. A careful
observation of daily life, of behaviour
and rituals as well as of the objects that
surround us, brings a surprising
playfulness to their work. This is an
approach which owes much to their
cultural roots in Japan. Furniture
changes from being one thing to become
another—a chest of drawers folds out to
become a low table, a shelf unit becomes
a stool. The mechanism of this
transformation is very much part of the
design. Like the ancient Japanese paper-
folding art of Origami, the movement
is simple and the result unexpected.
Above all, their work focuses on the
manner of using objects and suggests
new ways of interacting with them.
Objects should always give us pleasure—
not just when we look at them but also
when we use them.

Left:–
'Wire frame stool < shelf', 1998.
Made from the manufacturing process
used to make shopping trolleys and
hamster cages.

Far left:–
Wire frame chair prototype, 1998.   103

4

Please Touch the Criticism:
Design Exhibitions and Critical Design
in the UK, 1998–2001

The dark kitchen is in an exaggerated, almost post-apocalyptic, state of disarray. Washing up is stacked around the sink, a dirty towel languishes on a hook, on the table are a jar of dead tulips, a Nevvacold insulated teapot, and a Double-Bubble coffee mug in a quotidian inversion of a Dutch still-life painting, and in the background a broken microwave, an Argos kettle, and provisionally looped electrical wires are all in view. In the decrepit back yard are generic plastic chairs, a sprawling hose, a desultory scattering of weeds among the jettisoned flower pots, and peeling stucco on the back wall of the house.

These dystopian *mises en scène* of urban domestic life are the carefully staged backdrops for an upturned yellow bucket pendant lampshade by Michael Marriott, which hangs from the ceiling of the kitchen, and an orange chair-cum-shelf, by the Azumis, derived from the manufacturing process used to make shopping trolleys and hamster cages, in the yard.

The designer's studio kitchen and the anti-garden, as depicted in these photographs, have seemingly collected the aesthetic fallout of some off-stage explosion of all the values of good design pursued since the design reform era, and all the attributes of style codified in the 1980s. As such, they are telling tableaus of a late-1990s moment in recession-hued British design. Despite the best efforts of the 1980s tastemakers to help people choose "a better salad bowl," and of the critics to vivify design writing with pyrotechnic Wolfian prose or post-structuralist critique, the late 1990s saw a visceral reaction both against design, as evidenced in a generational "panic attack induced by good tastes in the kitchenware department of the Conran store,"— and, ultimately, the disintegration and silencing of the practice of written criticism.[1]

4.1
Spread from *Stealing Beauty* catalog, designed by Graphic Thought Facility, showing designer Michael Marriott's bucket pendant light hanging in his studio. Courtesy of Graphic Thought Facility.

4.2
Spread from *Stealing Beauty* catalog, designed by Graphic Thought Facility, showing the design duo, the Azumis' wire frame stool = shelf in their back garden. Courtesy of Graphic Thought Facility.

As the design boom of the mid-1980s fizzled out at the decade's close, designers were forced to renegotiate the identity of their profession in relation to the new realities of an economic recession, globalization, climate change, and anxieties surrounding the approaching millennium. Product design criticism, closely tied as it was to design's fortunes, also appeared to founder. Out of a weariness with the excesses of design celebrity culture, and a silence engendered by a sense of the futility of critical judgment—of language, even—there emerged in 1990s Britain two nonverbal alternative modes of product design criticism: the atmospheric exhibition and the designed object itself.

INTRODUCTION

This chapter examines the design exhibition as a means of conducting design criticism, and shows the ways in which it provided alternatives to, and even supplanted, the role of journalistic design criticism in late-1990s London. It contrasts two exhibitions: "powerhouse::uk" was a 1998 Department of Trade and Industry initiative, emblematic of New Labour's attempts to rebrand Britain in corporate terms, using design and creativity as nation-defining qualities as well as international political and economic tools. The other exhibition considered here, "Stealing Beauty: British Design Now," was held at the Institute of Contemporary Arts (ICA) in 1999, and collected the work of designers who used the "everyday" as conceptual and material inspiration and whose products were oriented toward an urban domestic setting, thus complicating any outwardly projected vision of a national identity based on design, and design as a national export.

In addition to analyzing the exhibition as a critical device, this chapter also examines the emergence of a genre of fictional furniture-appliance hybrids by Dunne & Raby, a design practice formed in the early 1990s, whose work was displayed in both the aforementioned exhibitions as well as many others of the period.[2] Dunne & Raby's work, which at the time they labeled "critical design," countered design's then-established role as a problem solver and a profit generator. Their invented products, destined for a near but undefined future, were characterized by their lack of obvious style or function and their intention to question social norms and help articulate anxieties, particularly those surrounding information technology and electronic products.

Both the exhibitions and the designed objects on display were produced against the backdrop of a declining goods trade and manufacturing industry in late-1990s Britain, and a concerted drive on the part of government-endorsed institutions to generate investment in British design. At the same time, the very notion of national identity was being undermined by the inexorable rise of digital technologies that seemed to intensify the effects of economic globalism and the sense of "a territorial contiguity of nations," creating what philosopher Paul Virilio termed "the telepresence of the era of globalization."[3]

Cool Britannia-flavored design

In 1997, after eighteen years of Conservative government, Britain elected a Labour government with the forty-three-year-old Tony Blair as prime minister. Blair had helped lead the party's dramatic repositioning: "New Labour," a term first used as a conference slogan in 1994 and cemented in a 1996 manifesto, *New Labour, New Life for Britain*, represented a shift in party values away from traditional tenets of socialism and trade unionism and toward more centrist policies and market economics.[4]

Design, which had become ideologically enmeshed with Thatcher's enterprise culture in the 1980s, was increasingly reframed in New Labour's political rhetoric as "creativity" and "innovation"—qualities perceived to be more encompassing than design, and more representative of the reassignment of economic value from traditional production-line industry to the market-dependent service sectors of banking, advertising, design consultancy, media, property, and retail. In their efforts to reposition Britain in the global knowledge-based economy, New Labour adopted the language of marketing, encouraged by their media-savvy director of communications, Alastair Campbell, and embarked on a national rebranding effort that came to be known as "Cool Britannia." This phrase, also a Ben & Jerry's ice-cream flavor featured on the cover of a 1996 issue of the American publication *Newsweek*, was used as a catchall for British creativity: "Britain has a new spring in its step. National success in creative industries like music, design and architecture has combined with steady economic growth to dispel much of the introversion and pessimism of recent decades. 'Cool Britannia' sets the pace in everything from food to fashion."[5]

The reality of British design as an industry was somewhat bleaker, with its weakened manufacturing base and a poor international image. The Design Council was dramatically cut and restructured in 1994, resulting in the closure of its Design Centre and regional offices. Government-funded institutions such as the British Council and the Crafts Council were also focusing their attention on creativity, rather than manufacturing, as a particularly British quality and a marketable export. Britain's state-endorsed design organizations were obviously concerned with British design's image abroad, and sought to counter a longstanding and entrenched governmental reliance on "heritage" as a national export with a more modern conception of Britain as "a global island, uniquely well placed to thrive in the more interconnected world of the next century."[6]

This notion of Britain derived from a report published in 1997 by the independent think-tank Demos.[7] Commissioned by the Design Council, written by Demos senior researcher Mark Leonard, and titled *Britain™: Renewing Our Identity*, the report recommended a rebranding of national identity which would be largely dependent on capitalization of home-grown creativity and design.[8] The upbeat views and the marketing language of this report quickly entered the lexicons of New Labour and design rhetoric of the period.

Taking creativity on the road

In art practice, curating gained currency in the mid-to-late 1990s, a period which Michael Brenson called "the curator's moment."[9] Paul O'Neill suggests that artists were turning to curation as a new means of generating debate in response to the silence of the art critic: "The ascendancy of the curatorial gesture in the 1990s also began to establish curating as a potential nexus for discussion, critique and debate, where the evacuated role of the critic in parallel cultural discourse was usurped by the neo-critical space of curating."[10]

Exhibitions, trade shows, and traveling showcases of contemporary British design proliferated in the 1990s, enabled by funding from sources such as the 1993 National Lottery Act, and impelled by a mission on the part of the government and government-funded institutions to promote British creative industries. International furniture trade shows were expanding with an increasing number of fringe exhibitions. For example, *Blueprint* magazine and the British Council staged a supplementary exhibition for the 1998 Salone del Mobile in Milan called "Zuppa Inglese." It consisted of filmed interviews with eight British designers and architects (including Dunne & Raby) and a set of eight customized traveling cases containing representations of each designer's creative influences. In London several new trade shows were initiated to provide commercial platforms for contemporary design, including 100% Design in 1995 and Designers Block in 1996. Museums such as the V&A exhibited contemporary design in their Design Now room and in large-scale temporary exhibitions such as 1991's "Visions of Japan." The Design Museum, launched in 1989, held exhibitions throughout the 1990s organized around themes such as sports, French design, or plastics, and retrospectives of designers such as Paul Smith, Philippe Starck, and David Mellor; and in 1994 initiated its Conran Foundation Collection, which presented selections of contemporary design. Designers also showed their own work in their own gallery spaces, like Tom Dixon's Space gallery, or like Droog, the Dutch design collective, who organized traveling exhibitions of their oeuvre.

The Department of Trade and Industry (DTI) was particularly active in the 1990s, arranging several exhibitions to promote British design abroad, and the BBC Design Awards program, launched in 1986, began in 1996 to be accompanied by exhibitions of its finalists, staged throughout the UK. City-specific design festivals, such as the Glasgow UK City of Architecture and Design 1999, directed by former *Blueprint* editor Deyan Sudjic, provided another opportunity for temporary exhibitions. Most monumentally, design was also included among the exhibits in the Millennium Experience, Britain's controversial and ultimately underperforming celebration of the new century, sited in Greenwich and open to the public during the year 2000.

Due to the nature of their funding and the missions of their organizing institutions, most of these exhibitions were promotional, providing little opportunity for critical reflection on the part of their curators. They presented variations on the theme of design and creativity as marketable assets in the political project of asserting a dynamically reconceived national identity.

PART ONE: THE DESIGN EXHIBITION IN LATE-1990S LONDON

"powerhouse::uk": inflated and babbling

One such promotional exhibition, and a highly visible example of New Labour's co-option of design and creativity under its "Cool Britannia" banner, came in the first few months of its administration. "powerhouse::uk" was an exhibition commissioned by the Department of Trade and Industry to encourage a global community to purchase British products and invest in British industry. Its opening was timed to coincide with the Second Asian Europe Summit (ASEM2), which was hosted in London on April 3 and 4, 1998. The summit was attended by heads of state and government from ten Asian and fifteen European nations, and the DTI, wanting to capitalize on the presence of business delegations and large media teams, saw this as "an opportunity to demonstrate, to an influential audience, how British creativity has led to world class products and services in design, fashion, technology, engineering and scientific research."[11]

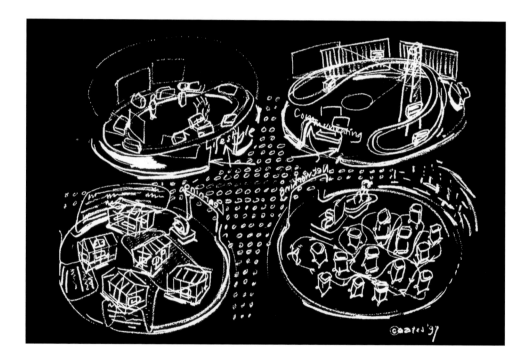

4.3
Sketch for "powerhouse::uk" by Nigel Coates, 1998, showing organization of exhibition contents into four pods. Courtesy of Nigel Coates.

"powerhouse::uk" took its name and much of its tone and terminology—which included such buzzwords as "hubs," "hybridity," "networking," "connectivity," and "innovation"—directly from the 1997 *Britain™* Demos report. Architecture critic Hugh Pearman, writing in the *Sunday Times*, observed the direct link between the report and the DTI exhibition, suggesting that architect Nigel Coates's involvement in both projects was partly responsible.

The architectural practice Branson Coates designed the £1 million exhibition space, a silver, four-drum inflatable structure, which was staged on Horse Guards Parade in Whitehall, and open to the public for two weeks.[12] Each sixteen-meter steel-framed dome could contain 300–400 people and was clad in silver-coated polyester PVC membrane, the pockets of which were puffed out by a low-power electric fan. The inflatableness of the architecture was not structural, therefore; it was a skin intended to attract attention. Being sited on Horse Guards Parade, the structure could not have foundations, so it was weighted down by concrete entrance ramps and electricity was provided through cables, which ran to a generator in Admiralty Arch. Branson Coates won the DTI's competition at the end of 1997 and were charged with designing, building, curating, managing, and deinstalling the exhibition in three and a half months.

Although it was Branson Coates who brought Claire Catterall on as curator, ultimately, Catterall's authorial role in the exhibition was subsumed by their architectural vision. The architects were not concerned with curation in the sense of telling a particular story through objects. Instead they designed a spectacular environment, in which the selected exhibits became absorbed into the very structures of the exhibition design. They wanted to convey a surface-level impression of British creativity, and had neither the inclination nor the time to probe beneath that surface to analyze the significance of specific examples. The exhibition was divided into four sections, one in each pod of the structure. In the "Communication" pod, which comprised graphic design, advertising, special effects, computer games, and film, examples of packaging design were used to construct a London cityscape with a St. Paul's Cathedral made from Conran's Bluebird wine boxes, book jackets, tins, and CD cases. Toy buses and taxis, customized by selected designers, whizzed around the packaging city on a Scalextric track. The "Lifestyle" pod, which encompassed industrial design, furniture, and fashion, featured a luggage carousel, which dipped and veered around the room conveying 31 open suitcases containing Manolo Blahnik stilettos, Paul Smith suits, Ron Arad stacking chairs, Tom Dixon "Jack" lamps, and Psion calculators, as if already packed and ready for export.

4.4
Exterior shot of "powerhouse::uk" inflatable exhibition structure located on Horse Guards Parade, London, designed by Branson Coates, 1998. Courtesy of Nigel Coates.

4.5
Interior shot of "powerhouse::uk" showing the packaging cityscape in the Communications section of the exhibition, designed by Branson Coates, 1998. Courtesy of Nigel Coates.

4.6
Working sketch of Communications section of
"powerhouse::uk" exhibition, designed by Branson
Coates, 1998, showing use of buzzwords and
floating phrases which Judith Williamson termed
"babble." Courtesy of Nigel Coates.

Architecture critic Giles Worsley observed: "Some architects reckon that if they have been asked to design an exhibition it is because their work is quite as interesting as anything on display. That was certainly true of Coates's 'powerhouse::uk', which was really no more than a trade show full of bio-crops and wackily inventive vacuum cleaners. The impact lay more in the totality of effect than in the individual objects."[13]

Nigel Coates was under no illusion that "powerhouse::uk" was anything more than a trade show; in fact he highlighted its commercial objectives, saying: "everything here is connected to business in some way."[14] And Trade and Industry Minister John Battle baldly identified the exhibition as an example of Britain "setting out its stall better."[15] But Ian Peters of the British Chambers of Commerce argued that what was really needed to improve the British economy was "a lower level of interest rate and a stable economy" and "long term investment," a line of policy thinking he saw as being resolutely ignored by New Labour's public message.

The extravagant ambition of "powerhouse::uk," and its emblematic role in New Labour's efforts to deploy creativity as a national branding tool, put it at the center of the gathering backlash against the "Cool Britannia" campaign. According to an article in *PR Week*, PR agencies were beginning to advise celebrities to disassociate themselves from "Cool Britannia."[16] Unsurprisingly, the balloon-like quality of the structure lent itself to charges of being full of "hot air" and, as such, a physical manifestation of empty political rhetoric.[17] Commentators on the exhibition were particularly distrustful of its use of "marketing jargon" and "US business school language."[18] Judith Williamson, who reviewed the exhibition for the graphic design journal *Eye*, took issue with its language—"babble," "blab," "meaningless chatter," and "self-congratulatory streams of dislocated words and circular messages," as she variously referred to it—and the ways in which such language did not speak of the actual creativity on display, but merely reflected the values of the politics that shaped it. She focused on the emptiness of the words and phrases that were projected on giant screens (intimate, rain, memory, work, laugh and hand me down, splash it all over, and west end girls): "The room was a babble of electronic messages made up largely of buzzwords and clichés. It was clear that they were meant to invoke a medley of British lifestyles and cultural trends. But what they invoked most of all was, appropriately, precisely the increasing bombardment of repetitive lifestyle verbiage that makes up much of British culture at present."[19] She saw the catalog to the exhibition, "packed as it was with buzz-words about creativity, innovation, mapping, diversity," as "the hard-copy counterpart of the digital babble in the show itself. For the most part, the babble *was* the show."[20]

Another target for criticism was the exhibition's evocation of a nation networked by intangible digital technology, which seemed out of touch with the still-fragmented, localized, and very tangible realities of the country's decrepit physical infrastructure. Architecture critic Jonathan Glancey wrote: "For many first-time visitors to Britain—including the business executives the Government wishes to woo—the impression here is one of garish carpets that disfigure the airport lounges, deregulated buses, clapped-out privatized railways, major roads in a permanent state of disrepair or being repaired, trashy 'vernacular' housing, people sleeping in doorways, overflowing rubbish bins."[21]

The overflowing rubbish bins of Britain, evoked by Glancey, were the point of departure for a very different exhibition, held the following spring at the ICA. "Stealing Beauty: British Design Now" portrayed what its curator, Claire Catterall, saw as a new sensibility evident in design and the way it was being practiced in late-1990s London, "a mixture of passion, beauty, rough edges, rawness," based on a relish for, rather than a repulsion toward, rubbish.[22]

"Stealing Beauty": a complete environment

The ICA, established as an alternative meeting space for artists, writers, and scientists, was not concerned with promoting British design, or indeed any worldview in particular. It was a broadly defined arts institution, staging avant-garde experimental performances and art exhibitions, partially publicly funded, but also heavily dependent on commercial sponsorship by brands such as Perrier-Jouët Champagne, in the case of "Stealing Beauty."[23] It operated in an interstitial space between government, culture, and commerce, and afforded Catterall—who was born in Malaysia, and already considered herself an outsider—a position beyond both the state- and commerce-driven demands on a design exhibition like "powerhouse::uk."

"Stealing Beauty" was hastily assembled in three months, with a small budget of £20,000 (compared to the £1 million spent on "powerhouse::uk" or the £250,000 spent on "Culture and Commerce," the Design Museum's inaugural 1989 exhibition), yet its influence extended well beyond its modest scope.[24] It was widely reviewed in the national press, lifestyle publications, and art and design magazines, generating at least 50 reviews and features thanks to formidable press outreach on the part of the ICA.[25]

The exhibition gathered the recent work of seven individual designers and nine design collectives, with some pieces specially commissioned. Most of the designers were British; the foreign-born ones were based in Britain. They were in their twenties and thirties, and had recently graduated. Some of the designers worked in the spaces they lived in, such as the three members of El Ultimo Grito, who lived and worked in a council flat in Peckham, and many of the objects they produced were small-scale, multifunctional and provisional fixes to domestic quandaries, or attempts to bring small moments of beauty into their low-rent, sparsely furnished living spaces. The work these designers contributed to the exhibition was concerned with the experience of living a transient, noncommittal, urban existence. It turned away from public issues of deregulated capitalism, environmental catastrophe, and globalization, and looked inward instead to issues of personal meaning; it functioned in a circumscribed sphere in which designers designed primarily for, and among, themselves.

a) Raiding the rubbish

Most of the exhibits were made from or inspired by found materials, or "things stolen from the landscape of our everyday lives," as Catterall put it in the exhibition catalog.[26] She saw the use of scavenged materials and the act of "urban hunting and gathering" as deliberate responses to the slick processed materials used by more established designers. The improvised and ad hoc approaches to design represented in "Stealing Beauty" also implicitly referenced the social degradation that occurs under capitalism, and the potential of engagement with the everyday as a liberating alternative to style-based conceptions of design.

By repurposing pieces of mundane detritus such as bus tickets, lottery numbers, and secondhand clothes, the designers featured in the exhibition celebrated the everyday as "an arena of authentic experience," as the literary theorist Rita Felski termed it.[27] The work can be seen as a delayed and largely depoliticized material manifestation of the ideal of an engagement with the everyday as a way to resist power structures, cut across class barriers, and problematize capitalism and society's infatuation with "the spectacle," as theorized by Henri Lefebvre and the Situationists of 1960s Paris.[28]

The exhibition's title, with its use of the term "beauty," was selected to suggest an urge to transcend the everyday, and to render the ordinary extraordinary.[29] The exhibition went through many name changes before "Stealing Beauty" was approved by ICA director Philip Dodd. "Nothing Out of the Ordinary," a title which evokes more closely Felski's "world leached of transcendence," was Catterall's preference, but Dodd considered that the word "ordinary had pejorative overtones."[30] Most of the work in the exhibition engaged with the everyday, however, not as a negative or residual state to be transcended or resisted, but rather as the expression of the small pleasures to be found in "repetition, home, and habit" (Felski's conception of three facets of the everyday), pleasures that could be embodied through the methods, circumstances, and materials of making, as well as consuming.[31]

The design firm El Ultimo Grito, who produced hybrid, multifunctional furnishings, exhibited their "Millwall Brick," a rolled-up newspaper secured with a piece of wire to make a coat hanger. Swedish fashion designer Ann-Sofie Back's garments were made from secondhand clothes, which she reconstructed with new additions of plastic bags, safety pins, and color from felt-tip pens, while George Badele's Stalagmite lanterns deployed stacked rolls of masking tape. The furniture designer Michael Marriott repurposed an inverted bucket as a pendant lampshade and a sign found at London's disused Aldwych underground station as a table. In "Furniture for people without gardens," he constructed a temporary living space from plywood frame and plastic sheeting as walls, with pieces of furniture designed to support combinations of flower vases. As such, these pieces fitted with an impulse in design of the 1990s to reject the perceived excesses of the 1980s and return to minimalist or neofunctional forms, humble materials, and the designer's soberer public presence. "Humility is an inevitable step in the cleansing process that has been taking place in design," observed design historian Penny Sparke.[32]

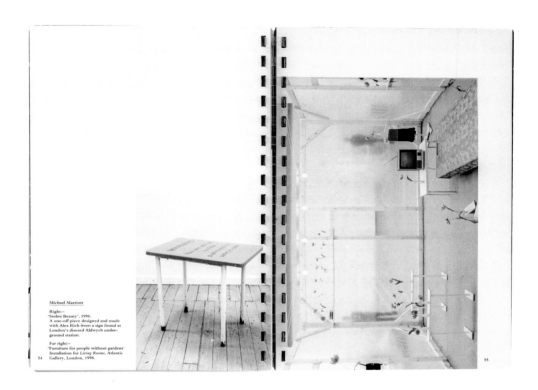

Michael Marriott

Right:–
'Stolen Beauty', 1998.
A one-off piece designed and made
with Alex Rich from a sign found at
London's disused Aldwych under-
ground station.

Far right:–
'Furniture for people without gardens'
Installation for *Living Rooms*, Atlantis
Gallery, London, 1998.

54

55

4.7
Spread from *Stealing Beauty* catalog designed by
Graphic Thought Facility, showing Michael
Marriott's "Furniture for people without gardens,"
a temporary living space made from plywood
frame and plastic sheeting, with pieces of furniture designed to support combinations of flower
vases. Courtesy of Graphic Thought Facility.

In a special section of the October 1997 issue of *Blueprint* titled "Product Overload," contributing editor Rick Poynor wrote that the contemporary shopping experience involved "too much variety. Too much duplication. Too many choices to make that have nothing to do with need. Too much fantasy. Too much stuff."[33] This condition presented a "central dilemma" for designers of consumer goods: whatever they produced—however well-intentioned, thoughtful, or alluring—would simply contribute to the "gigantic over-production of things."[34] The design critic used to be able to mitigate the situation by pre-sifting the stuff and helping people make informed choices. But a decade or more later, Poynor observed, "design-watchers" appeared to be paralyzed and were leaving TV, the newspapers, and the shelter magazine *Wallpaper*, "a buy-it-all bible of 'urban modernism,'" to tell the dominant story of design—as consumption, business opportunity, and status symbol. "An alternative vision of design, not dedicated to consumerist over-production, has all but disappeared within design itself as well as the press," Poynor averred.[35]

Concerns over climate change, implicit in Poynor's comments about overproduction, also contributed to the designer's dilemma. Curator and critic John Thackara summed up the impotency felt in the late 1990s: "For 30 years scientists, think tanks, and global summits, have measured and analysed the 'environment'. ... They've produced a stream of such ghastly projections that many people have been de-motivated by deep eco-gloom. ... The 'eco-problem' leaves us with guilt, denial, despair, or a combination of all three."[36] By logical extension, he and others implied, a green designer is one who designs nothing at all.

Some designers responded to this stymieing of the ostensible goals of their profession by retreating from the extravagances of 1980s design and focusing instead on modest incursions into the domestic environment that used recycled or cheap materials, and ready-made production processes. Paul Neale, a founding partner of Graphic Thought Facility, who designed the exhibition graphics and the catalog for the "Stealing Beauty" exhibition, recalled how "working with everyday undervalued materials," "optimising small opportunities," and using "modesty as a component" of practice were ways of reacting against the "style-led design of the late 1980s."[37] The exhibition wall panels were engraved on laminate, and the catalog was comb-bound with cheap wood-free paper stocks and packaging box board to evoke a utilitarian commercial brochure, in distinct reaction to the refined production quality of a more typical glossy art catalog.

In a review of the 1993 Royal College of Art degree show, David Redhead observed: "Everywhere there was modular, minimal and everyday furniture made of easily assembled, eco-friendly materials."[38] The students' work, Redhead argued, was symptomatic of "a broader European shift away from self-indulgence and flamboyance and towards self-denial and restraint which Italian critics have already christened New Functionalism." The furniture designer Jasper Morrison, who had been working in this austere mode even during the 1980s, told Redhead he "believed that designers have once again begun to think about the 'contextual value' of an object to its user and to restate fundamentals—usefulness,

4.8
Laminate wall graphics for "Stealing Beauty"
exhibition designed by Graphic Thought Facility.
Courtesy of Graphic Thought Facility.

4.9
Cover of *Stealing Beauty* catalog designed by
Graphic Thought Facility, featuring the cutouts
produced during the catalog's comb binding
process, photographed by Angela Moore.
Courtesy of Graphic Thought Facility.

PLEASE TOUCH THE CRITICISM

STEALING
BEAUTY
BRITISH DESIGN NOW
AN ICA EXHIBITION IN ASSOCIATION WITH
PERRIER-JOUËT CHAMPAGNE

longevity, and ordinariness—that were squeezed off the agenda in the rush for self-expression."[39] Some designers reasoned that the more connected someone felt to a product, the longer they would probably keep it, and the less damage it would do to the environment, and so they sought ways to ignite emotional responses to their work. In 1998 the Eternally Yours Foundation, a Dutch product think-tank, published *Eternally Yours: Visions on Product Endurance*, a book that made the claim that green thinking needed to focus on how to persuade people to keep their products for longer, through the use of well-built hardware, updatable software, and by making them lovable. Taking the Eternally Yours project as one of its reference points, "Stealing Beauty" drew attention to the emotional resonance of designed objects—how objects could activate and embody the memories of both their designers and their users.

As Catterall remarked: "I think the designers wanted to put out something familiar and something you could respond to on an emotional level. They wanted to show that design wasn't a global monster that has no integrity, personality or intimacy. Design is driven by need but also by emotional need. It was a turning point, really, when we realized that design could really make you feel different, that it could provide comfort."[40]

b) Making do

Technological developments such as desktop publishing, Computer Aided Design, and Computer Aided Milling altered the way designers worked in the 1990s, allowing for smaller studios and more rapid prototyping, while increasingly computerized production processes and larger scales of production led to a more risk-averse manufacturing climate. Miniaturized electronics enabled by the microprocessor chip necessitated acts of translation on the part of designers, who were asked to create readable interfaces to allow the operation of appliances where the mechanisms were not easily understood. The abstract qualities of new technologies, such as lightness, transparency, transformability, and elasticity, gave rise to anxieties over the dematerialization of objects.

Most of the designers featured in "Stealing Beauty" made a virtue of their enforced role as postindustrial designers-as-makers. Making the things themselves, and showing the public how they could do so too, was a response to their lack of access to Italian manufacturers like Cappellini and Moroso, who tended to work with well-established names. The neutral authorial voice of these designers, the self-consciously provisional nature of their "make-do" solutions, and their dependence on default shapes and production processes and found materials were partly a reaction to the flamboyant stylistic flourishes of many designers in the public eye at the time, such as Philippe Starck, Ron Arad, Marc Newson, and Frank Gehry. Quality, craftsmanship, and signature styles were beside the point in chairs made of plywood and army blankets; these were anti-luxury statements.

Some used existing manufacturing processes but subverted their intended use for their own purpose. The architecture firm 24/seven, for example, appropriated Robin Day's

1964 polypropylene chair and changed the production process to turn the normally brightly colored seat to monochrome and, for their bar design in the ICA café, they specified the fireclay used by Staffordshire ceramics firm Armitage Shanks for toilets and urinals.

Another way designers responded to their collective professional guilt about the perceived overexposure of design and its celebrities was to work collaboratively, and to conceive of themselves less as authors of complete works, more as facilitators of social interaction and "co-design." They created half-finished and ambiguous products, which needed to be completed and interpreted by their users in the sense of "labour-to-be," as art theorist Nicolas Bourriaud put it.[41]

The Dutch designer Tord Boontje exhibited his "Rough and Ready" furniture and lighting made from materials that could be "found in the street and on building sites," such as softwood, plywood, chip board, screws, army blankets, plastic sheeting, secondhand fluorescent tubes, and metallic tape. Boontje provided exhibition-goers with instructions and a list of materials so they could make their own chairs at home. "The unconcluded appearance of the pieces makes them feel as if they are subject to change," wrote Boontje in notes accompanying his exhibits.[42]

This kind of work was similar to socially collaborative art, practiced by artists such as Martha Rosler, Carsten Holler, Jeremy Deller, and Rirkrit Tiravanija and typified, in Claire Bishop's words, by its "striving to collapse the distinction between performer and audience, professional and amateur, production and reception."[43] Like the artists interested in participation, designers found the space of an exhibition to be an ideal testing ground for their work. Most young British designers' work was unlikely to be put into production, while exhibitions provided them with a rare opportunity to introduce their work to the public. The participation that designers like Boontje offered with his kits of parts was limited, however. Users, in the role of deferred assembly laborers, followed a prescribed set of instructions; there was little room for creative input on their part.

c) The anti-lifestyle style

"Stealing Beauty"'s most explicit critique was directed against a consumerist culture and a fetishization of design and lifestyle that had developed in the 1980s. As Catterall wrote in her exhibition catalog essay, "'Stealing Beauty' is partly a reaction against the current saturation of the media by design—all those books, magazines and TV programmes which offer instant access to 'stylish living,' dispensing advice on how to 'get the look' and performing makeovers on our homes."[44] Catterall saw lifestyle being used by the "style mafia" as a panacea for a society in crisis, or at least a state of malaise, one marked by "feelings of deep insecurity, in ourselves and our role in life, and in the machinations of a world where even the axes of time, space and reality are disintegrating."[45]

The "style mafia" Catterall invoked were represented most literally in the pages of *Wallpaper*, a magazine launched in London in 1996 by journalist and entrepreneur Tyler

4.10
Spread from *Stealing Beauty* catalog designed by Graphic Thought Facility, showing Tord Boontje's building instructions for his Rough and Ready chair. Courtesy of Graphic Thought Facility.

PLEASE TOUCH THE CRITICISM

Brûlé, which enfolded design coverage with travel, fashion, and lifestyle. Its characterization of design as a necessary component of the kind of style-conscious, jet-setting way of life that Brûlé espoused was extremely successful in terms of publishing strategy, yet easily lampooned and quickly rejected by a younger generation of designers who found the glamorous lifestyle depicted in its pages out of touch with the concerns of their everyday lives. When *Wallpaper* wrote of the mission of "Stealing Beauty" ("the ICA's snappy new exhibition") as being "to reverse the 90s obsession for the sleek, the shiny, and the sanitized," they must have recognized themselves in its wording, for they added: "though we have always maintained that a little of what you fancy does you good."[46] The title of the magazine became a popular adjective to describe a genre of injection-molded furniture, products, or "blobjects" (dictated by the spline curve allowed for by computer-aided design technology) and sinuously surfaced interiors prevalent in 1990s Britain. As Richard Benson put it in *The Face*, "A lot of people are sick of super-slick, Wallpaper-esque bars and furniture, and all that taupe and curvy-cornered stuff is looking suspiciously like angular matt black things did around '89."[47]

A specific target for the "Stealing Beauty" designers' angst was Terence Conran and his propagation of stylish living through design, which had spread so vigorously in the 1980s. In answer to the question "What is your worst design memory?" the architectural practice FAT (Fashion Architecture Taste) had responded: "A panic attack induced by good tastes in the kitchenware department of the Conran store."[48]

The anti-lifestyle theme of the exhibition was embodied most directly in the photographs included in the "Stealing Beauty" catalog. Objects were photographed in exaggeratedly banal and messy environments, as discussed at the beginning of this chapter. Ann-Sofie Back's clothing was depicted on deliberately unglamorous models, with partial body shots pieced together in mismatched sections, like a game of Exquisite Corpse. With their calculated nonchalance, these anti-glamor shots (redolent of Wolfgang Tillmans's still-life photography of the detritus of everyday life) were clearly styled just as much as those in the design magazines and showroom catalogs that they sought to counter.[49]

Through critiquing lifestyle culture, the designers replaced it with another anti-lifestyle aesthetic, which itself would become increasingly commodified in the ensuing years. Celebrating the imperfect make-do approach to production became a stereotyped practice. As the critic Nick Currie observed of the exhibition, it "failed to avoid the post-materialist paradox: attempts to snub status-seeking quickly become new claims to status."[50] And Giles Reid, writing in *Object* magazine, averred: "'Stealing Beauty' didn't blur boundaries between high and low design, rampant materialism and gritty realism. It only entrenched a new aesthetic range of appreciation to maintain an elitist hold on the proceedings."[51]

And yet, at that moment in the late 1990s, the work of the designers on view in "Stealing Beauty" did seem to present an alternative to the slick, lifestyle-oriented notion of design that dominated retail and design media. Catterall explained: "If design caters only for those who can afford it, who subscribe to a certain ideal and approach to life, what is left for those who cannot aspire to such lofty heights or simply don't want to? In this light, the work can be seen to have a political and social resonance only because it responds so directly to the circumstances of its need, conception, production and, ultimately, its consumption and use."[52]

The exhibition would inevitably be caught up in a process of mainstream appropriation and commodification whose speed had increased to the point where it was happening in parallel to the production of the work itself. The exhibition could never exist completely outside the predominant strain of design discourse, but in seeking to present a strand of contemporary design still in formation, "Stealing Beauty" attempted to offer a counterstatement. As Paul O'Neill observed of art exhibitions with similar ambitions, using cultural critic Raymond Williams's conception of dominant, residual, and emergent cultural moments, "emergent cultural innovation comprises new practices that produce new meanings, values and kinds of relationships. Emergence is thus not the mere appearance of novelty: it is the site of dialectical opposition to the dominant—the promise of overcoming, transgressing, evading, renegotiating or bypassing the dominant—and not simply delivering more of the same under the blandishments of the 'new'."[53]

While all the designers featured in "Stealing Beauty" made use of free or inexpensive materials and processes, the resulting work was largely inaccessible; the products and proposals were limited editions, prototypes, and one-offs. As Gareth Williams, assistant curator in the Furniture Department at the V&A, pointed out, "Many people expect an ICA show to be transgressive just because of the venue. I suspect design may suffer more than art in this environment, as we understand art to be made for these rarefied places. Design on the other hand is still primarily to be used in the real world. Design in a gallery can appear precious and even pretentious, not because it is intended to be so, but because it is out of context."[54]

In the case of "Stealing Beauty," however, the exhibition space at the ICA *was* the intended context for the design. There was no real world in which the objects once existed and from which they were subsequently decontextualized, and in that sense the show was a critique of young designers' lack of access to manufacturing and distribution deals. (The ICA bookstore sold the pieces that could be produced in multiples, and took 50 percent of the retail price.)[55] Most of the work on display was produced specifically for the exhibition, and the pieces were carefully juxtaposed to create environments specific to the gallery. 6876's jackets in "pavement grey" and "steel blue" hung above George Badele's two-tone floorboards in which layers of paint had been exposed by the wear of feet. Tord Boontje and Michael Marriott provided the furniture and lighting; Bump supplied the cups and plates, which were labeled with insults suitable for plate-throwing arguments. All the appurtenances of the Millennial London designer's domestic interior were represented in this composite portrait of design in thrall to the everyday.

4.11
Interior shot of "Stealing Beauty" exhibition, showing 6876's jackets in "pavement grey" and "steel blue" hung above George Badele's two-tone floorboards, in which layers of paint had been exposed by the wear of feet. Exhibition designed by Urban Salon. Courtesy of Urban Salon.

4.12
Interior shot of "Stealing Beauty" exhibition, showing the Azumis' wire frame chairs made at a supermarket trolley production plant, and in the background, plates with insult transfers for throwing in domestic arguments by Bump. Exhibition designed by Urban Salon. Courtesy of Urban Salon.

4.13
Interior shot of "Stealing
Beauty" exhibition, showing
FAT's woodland of silver
birch trees. Exhibition designed
by Urban Salon. Courtesy of
Urban Salon.

4.14
Interior shot of "Stealing
Beauty" exhibition, showing the
Light Surgeon's audiovisual
installation. Exhibition designed
by Urban Salon. Courtesy of
Urban Salon.

PLEASE TOUCH THE CRITICISM

d) Objects in conversation

In marked contrast to Stephen Bayley's exhibition-as-magazine-article approach to curating at The Boilerhouse, discussed in chapter 3, Catterall made minimal use of wall texts and captions in the exhibition, using them only to orientate the visitor rather than to explain the objects on display. Catterall was interested in curating as a largely nonverbal practice, which was less about illustrating a prewritten essay with objects than it was about "weighing things up against each other, and seeing how they react to each other."[56] She preferred to exercise her curatorial judgment by editing out "dead wood"; using juxtaposition to create "conversation" between objects; accumulating multiples for rhetorical effect; "precisely" positioning objects; and creating atmosphere through a constructed all-encompassing environment.[57] In notes for the exhibition, she wrote: "the few successful design shows are more like art installations—communicating something through the very space they occupy. It's time to change the form and format of design exhibitions—so that they engage, challenge, and provoke."[58]

The exhibition, designed by Urban Salon, delineated each designer's equally sized space with a colored strip, which extended across the gallery floor, up the walls, and into the corridor outside. The work was displayed on the floor, hanging from the ceiling, and leaning against walls. FAT used a mirror on the ceiling of their allotment-like strip to extend the height of their forest of silver birch trees. Exaggerated contrasts in lighting were used for dramatic effect, and various floor textures were employed throughout (maintenance instructions for the exhibition note that "Dunne & Raby's grass should be watered every day").[59] The stairway was flyposted with British Creative Decay's screen-printed images of anti-flyposting devices. In the upper gallery, video jockeys The Light Surgeons installed an immersive environment of projected images and video footage, which evoked one of the lightshows they created for clubs. Sounds such as those emanating from Dunne & Raby's talking plant labels, "Rustling Branch" and "Cricket Box," created by experimental musician Jayne Roderick, provided an aural backdrop for the work.

Catterall eschewed the use of plinths and vitrines. Her primary inspiration for curating in this sense-evoking and atmosphere-producing manner was the "Bodyworks" exhibition by Japanese fashion designer Issey Miyake, which originated in Toyko in 1983 and was shown at The Boilerhouse in 1985. Miyake displayed his work on custom-made black silicon mannequins hanging from the ceiling. Catterall enthusiastically recalls of the show: "it was a complete environment, it was about the inside of [Miyake's] head more than anything—no really wordy captions—but a whole environment, with torsos bouncing up and down. And it made you feel really fantastic. And in a way I think that's what the 'Stealing Beauty' exhibition tried to do—rather than putting an object on the plinth and just telling you that 'this is this' and 'this is about this.' You were kind of meant to go into the exhibition and feel it. Or taste it, as the case may be."[60]

e) A species of quiet criticism

"Stealing Beauty"'s quiet introspection contrasted emphatically with the bombast and "babble" of the "powerhouse::uk" exhibition which had taken place the previous spring. Both attempted to materialize the nebulous concept of British creativity, and represented a contemporary moment in time, but while "Stealing Beauty" was a ruminative exhibition, carefully contained beyond the fray of the marketplace in an independent art gallery, "powerhouse::uk"—from its macho name and its showy architecture to its alien-like landing in the middle of Horse Guards Parade—was intended to seduce a very particular audience of Asian businessmen and politicians. The form of Branson Coates's circular exhibition space was literally inflated, its ambition metaphorically so.

Art critic Hal Foster, reflecting on the "inflated" condition of design of the period, wrote of the way in which prices for design as a service (branding) and as an object (collectible pieces) were inflated in a contemporary situation in which there is no "running room" for culture. Everything is folded back into "the near-total system of contemporary consumerism."[61] Charles Leadbeater, author and advisor to Tony Blair, commented a couple of years later: "We are all in the thin air business these days ... most people in advanced economies produce nothing that can be weighed: communications, software, advertising, financial services. They trade, write, design, talk, spin, and create; rarely do they make anything."[62]

"Stealing Beauty," by contrast, was grounded by its focus on physical objects and the process of making, albeit a limited conception of manufacture. Furthermore, its comparative distance from the concerns of commerce enabled it to experiment and take a more irreverent stance. In the context of the ICA the exhibition performed as a space, not for business-focused discussion, but for contemplation of more poetic themes—epiphany, even.

The extent to which "Stealing Beauty" could achieve any significant critical distance toward its subject matter might have been compromised by its very format as an exhibition. According to Bruce Ferguson, exhibitions are always framed by institutional and commercial concerns, and will always perform ideologically: "Exhibitions are ... contemporary forms of rhetoric, complex expressions of persuasion, whose strategies aim to produce a prescribed set of values and social relations for their audiences. As such exhibitions are subjective political tools, as well as being modern ritual settings, which uphold identities (artists, national, subcultural, international, gender-or-race specific, avant-garde, regional, global, geopolitical etc.); they are to be understood as institutional 'utterances' within a larger culture industry."[63]

Yet the form of "Stealing Beauty" was more atmospheric than rhetorical. As an institutional "utterance," this exhibition was taciturn. It refused the model of exhibition-as-text and instead used the entire environment of the exhibition to convey its ideas. Catterall also rejected the art-historical tendency to label groupings of work as a "movement," preferring instead to characterize her selections as examples of a "mood and an energy."[64] Using minimal explanatory captions and, in the catalog, letting the designers speak for themselves through their response to questionnaires, "Stealing Beauty" left viewers space to elicit meaning or to remain confused by what they saw.

PLEASE TOUCH THE CRITICISM

"Stealing Beauty" critiqued the state of manufacturing, other designers, design retailers, the lifestyle press, and unthinking consumption. But it also critiqued the apparatuses of criticism through its nonlinear, non-narrative, format. As a piece of design criticism, "Stealing Beauty" relied on the palpable tensions and correspondences between featured objects to stimulate discussion about design in late-1990s London. "I'm hoping 'Stealing Beauty' will spark a renaissance of design shows which provide a platform for debate," Catterall told *Design Week*.[65]

4.15
Sketch for laminate sign of portrait of Dunne &
Raby and Michael Anastassiades, designed
by Graphic Thought Facility. Courtesy of Graphic
Thought Facility.

One measure of the exhibition's lasting effects can be seen in the work of Dunne & Raby, who at the time were formulating their own ideas about criticism. Their work, exhibited both in "Stealing Beauty" and in "powerhouse::uk," was positioned at the intersections of art and design, and of industry and academia. They explored the idea that criticism could be embodied in products and speculative proposals, and provide a viable alternative to the role of the journalistic design critic during the late 1990s. As a result, their work provides the most instructive example of the continued discussion of ideas presented in "Stealing Beauty," and a new trajectory beginning to move away from criticism as the design journalist's purview.

Dunne & Raby point to "Stealing Beauty" as a "pivotal" moment in their practice and a means of meeting a network of like-minded designers. They identified in particular with Alex Rich, El Ultimo Grito, and Michael Marriott, and went on to rent a studio in the same building as FAT. Dunne reflected: "up to that point we felt quite isolated, really like outsiders. 'Stealing Beauty' definitely made us feel like there were other designers who were doing really interesting work, and who we felt an affinity with, and kept in touch ever since."[66]

PART TWO: THE DESIGNED OBJECT AS CRITICISM

Products with a point of view

Although Dunne & Raby have relabeled their practice several times since that period, and currently do not use the term "critical design," in the mid–late 1990s they did use it to describe electronic product design's potential as criticism. "We view design as a form of criticism," they wrote, "where design proposals represent not utopian dreams or didactic blueprints, but simply a point of view."[67] They saw their work as a challenge to manufacturers and users "to question products through products."[68]

Their work also represented a challenge to design criticism as it was conventionally conducted, since they wanted to reposition criticism from its location in the media to a potentially more direct location within the design object. Dunne wrote: "design, too, has much to contribute as a form of social commentary, stimulating discussion and debate amongst designers, industry, and the public about the quality of our electronically mediated life."[69]

Dunne & Raby believed that electronic products could be designed to resist their passive consumption and unthinking acceptance and, as Dunne put it, "facilitate sociological awareness, reflective, and critical involvement."[70] In this conception, the act of criticism would be performed in a nonverbal dialogue between designers and users, thus deemphasizing the more established role of the critic as a skilled interpreter or translator between these two constituencies. In Dunne's conception of "critical involvement," the kinds of questions usually asked of an object by a critic would be embodied in the product itself, to be accessed by the user. The ways in which such products were presented assumed increased significance, therefore, and Dunne & Raby confronted the continuing

PLEASE TOUCH THE CRITICISM

challenge of how to create the conditions in which such questions could be made specific, or even legible.

In developing this brand of criticism, Dunne & Raby did not want to prescribe or moralize; they turned to the genres of film and fiction, seeking a mode of address that was "gentle and slightly subdued."[71] Dunne wrote in *Blueprint* magazine: "Industrial design's position at the heart of consumer culture (after all, it is fuelled by the capitalist system) could be subverted for more socially beneficial ends by enriching our experiences. It could provide a unique aesthetic language that engages the viewer in ways a film might, without being utopian or prescriptive."[72]

Part of Dunne & Raby's technique was to play with time, framing their work in the future subjunctive tense so that it could express various states of unreality such as desire, emotion, possibility, judgment, or action that has not yet occurred. Yet using that mode did not mean that they conceived of their objects as futuristic products; it was better that the future their objects spoke of was close at hand and, better still, one that could "sit uncomfortably alongside the now."[73] Their objects operated in fictive social scenarios set in a near future, or parallel present, in order to amplify current anxieties and practices. They blended critique of the present with projection into a hypothetical, prototyped future, exposing the mechanisms by which cultural values are made, and showing that it was still possible to reshape that future and those values.

Anthony Dunne and Fiona Raby met while studying Industrial Design and Architecture, respectively, at the Royal College of Art. Upon their graduation in 1988, the couple relocated to Tokyo, excited by the possibilities of a country with such a high rate of technological change, and disillusioned with the state of design manufacturing, the lack of advanced design research, and the drudgery required of a young architect in Britain. In Tokyo Raby worked with the experimental architect Kei'ichi Irie, and Dunne worked in Sony Corporation's Design Center. Here he created the prototype for Noiseman, a subversion of the Walkman, which recorded street sounds and distorted them to create an abstract ambient soundscape, thus reestablishing a link, albeit a transfigured one, between the Walkman wearer and the city he or she moves through.

When they returned to London, the couple established a collaborative practice. Their experience in Japan was pivotal, both through the contacts they made and through what they absorbed about the relationship between society and technology. "Tokyo is a city immersed in a sea of signs," wrote Dunne upon his return. "Every available surface is used to transmit information; clothes, objects, buildings all become screens, terminals for a vast information machine."[74]

One of their first clients was the Japanese architect Toyo Ito, who had been commissioned to create a thematic section of the "Visions of Japan" exhibition at the V&A in 1991. He had seen Dunne's Noiseman in Tokyo, and asked the couple to work with him. Ito created a "Dreams Room," in which hundreds of video clips of processed computer imagery and scenes from Tokyo life were projected onto the floor and walls. Dunne & Raby contributed a set of "media terminals" through which they addressed such questions as user-unfriendliness and the notion of data stored in spaces rather than in objects. Through this set of objects created for Ito, Dunne & Raby began to work out a design philosophy dedicated to revealing invisible aspects of the environment, such as electromagnetic fields.

The gallery space would provide Dunne & Raby with a public sphere for their ideas throughout the 1990s, but its location outside of everyday life troubled them. They considered themselves designers, not artists, and wanted to find ways to connect their work to lived contexts. Even when they got to turn an old TV salesroom in the Elephant and Castle Shopping Centre into a de-electrification center (as part of artists Rebecca Nesbit and Maria Lind's Salon 3 project), they were self-conscious about its art-world framing. In 1997, interviewed by *Blueprint*, Dunne commented: "we want to steer this debate away from a purely fine art context. Having our work shown at the Saatchi Gallery would be good. But being shown at Dixon's would be much better."[75]

Over the years the duo would continue to wrestle with the way their work was presented. They concluded that working in an academic environment provided them with the most freedom and potential, although even within this field they were keen to forge a new kind of practice fed from a continual exchange between research, making, teaching, lecturing at conferences, and exhibiting.[76]

Anthony Dunne's Hertzian Tales

Dunne & Raby joined a research group, which became known as the Critical Design Unit, funded by the Californian technology incubator Interval Research Corporation, and located in the RCA's Computer Related Design department. Raby recalled that the Critical Design Unit provided them with a kind of shelter to work through their frustrations with the design industry and a confusion over where their own work could exist if they rejected commercial design: "We used to joke that the Critical Design Unit was like a refugee camp for architects who didn't want to do architecture, product designers who didn't want to do products, and graphic designers who didn't want to do graphics."[77] Dunne enrolled as a PhD student and embarked on a six-year research project, published in 1999 as a book under the RCA/CRD imprint, titled *Hertzian Tales: Electronic Products, Aesthetic Experience and Critical Design*. The essays in the book explore historical precedents (particularly the work and thinking of Andrea Branzi from the 1960s and 1970s, Daniel Weil from the 1980s, and Ezio Manzini in the early 1990s), the work of peers in art, design, architecture, and literature (especially the instruments, projections, and vehicles of Polish-born industrial designer and director of the Interrogative Design Group at the Massachusetts Institute of Technology Krzysztof Wodiczko), and the thinking of philosophers (especially Jean Baudrillard).

Product semantics, an approach to design developed at the Hochschule für Gestaltung at Ulm in the 1960s, came to fruition in the early 1990s, and led to a focus among product designers on information displays, graphic elements, and the form, shape, and texture of a product as ways to indicate its function. *Hertzian Tales* critiqued this approach; Dunne dismissed what he saw as the prevailing emphasis on the optimization of the technical and semiotic functionality of products, arguing that most product categories have reached a watershed in terms of technical performance. He focused instead on a product's potential to provoke what he termed "psychosocial narratives," and to embody "inhuman factors" and "post-optimal aesthetics."

The book included a section in which he documented and reflected on five of his projects: each a variation on radio technology or, as Dunne put it, "an interface between the electromagnetic environment of hertzian space and people."[78] "Electroclimates" used a pillow-like PVC inflatable casing to contain a wideband radio scanner and a horizontally positioned LCD screen in a fluorescent polycarbonate box. It was created to be a kind of barometer of ambient electromagnetic radiation, that it converted into abstract sounds and pulsing patterns, which could be discerned by placing your head on the pillow. It was exhibited at the RCA exhibition "Monitor as Material" in 1996, but as an object, its commentary on the problematic interface between public and private space remained mysterious to the exhibition-goers. Dunne decided it needed contextualizing and so he, Dan Sellars, and Raby shot a pseudo-documentary video which depicted an elderly lady interacting with the pillow in her home, surrounded by doilies, teacups, and a copy of the *Sun*. This made the pillow's story much clearer, and drew an audience into shared speculation on its meaning; thereafter, Dunne was careful to present his objects in use through staged photographs or videos.

Another of the "Hertzian Tales" projects used changing color fields and sounds to visualize the intensity of electromagnetic leakage, or what he characterized in softly poetic terms as "dreams," from domestic consumer appliances like televisions, computers, babycoms, and fax machines. By contrast, "Thief of Affections" conjured a more perverse intentionality on the part of the object which could surreptitiously "grope" a victim's heart—via their pacemaker. When activated, a radio scanner concealed in a flesh-colored prosthesis resembling a riding crop or police truncheon would search for pacemaker frequencies in the vicinity, lock onto a close signal, and convert the frequencies into vaguely erotic audible sounds.

"Faraday Chair" took the form of a simple transparent box on legs, like a vitrine, inside which someone would be protected from EM. Its name referenced the contraption invented in 1836 by Michael Faraday in which a cage protects the occupant from electromagnetic radiation or radio waves by using conductive material to distribute the electric charges, canceling those within. Dunne's Faraday tank was not quite long enough for the person to lie outstretched, nor comfortable enough for relaxation, and so the supposed luxury and repose of a pure electronic radiation-free space was subverted by the inhabitant's awkward and vulnerable-seeming posture, like a baby in an incubator.

Dunne's "Hertzian Tales" were conceived of as stories in which objects figured as characters, props, plot devices, and atmospheres, and through which different values (spying, thieving, hiding) could be considered necessary means for survival in an increasingly electromagnetically radiated environment. Dunne was interested in shifting the focus away from the skin and interface of an electronic product and toward the kinds of psychological experiences that it could stimulate.

In a paper Dunne and fellow RCA researcher Alex Seago presented in 1996, titled "New Methodologies in Art and Design Research: The Object as Discourse," they attempted to validate the role of a designer as researcher within academia with the coinage "action research by project."[79] They wrote: "Dunne's work offers a positive and radical model of the action researcher in design as a critical interpreter of design processes and their relationship to culture and society rather than a skilled applied technician preoccupied by the minutiae of industrial production or a slick but intellectually shallow semiotician."[80] This was at a time when the value of applied academic research was being tested and contested in the academic sphere. Dunne's work, he and Seago suggested, provided a model of systematic research containing explicit data and reproducible methodologies: "the electronic object produced as the studio section of a doctorate is still 'design' but in the sense of a 'material thesis' in which the object itself becomes a physical critique."[81]

Exhibition as "reporting space"

Throughout the 1990s Dunne & Raby were invited to participate in design exhibitions in Britain and around the world.[82] Increasingly they considered an exhibition to be a "reporting space" into which they could bring a project, in order to gauge the public's reactions, and then incorporate those reactions in the project's development. The exhibition therefore began to function for Dunne & Raby as a part of the design process, and a space for testing their critical ideas.[83] And despite fruitful discussions with artists concerned with similar issues (such as Liam Gillick and Dominique Gonzalez-Foerster at "Le Labyrinthe moral," Le Consortium contemporary art space, Dijon, France, 1995), they also decided to move away from the art world: "Around the end of the 1990s, we said, 'No, we want to contribute to the design discourse, and be designers even if we don't fit in.'"[84]

In addition to using the gallery space as a "test site" for their work, in the late 1990s Dunne & Raby made increasing use of heavily stylized videos of fictional scenarios to present it. The scenarios they invented to frame their objects took place in alien, depthless worlds that, through their aesthetic unfamiliarity, are also morally disorienting. Dunne & Raby thought these videos were able to "focus the viewer's attention on the space between the experience of looking at the work and prospect of using it."[85] They still sought a situation where people might actually engage with their objects in their own homes.

Placebo Project

With "Placebo Project," a body of work created in 2001, Dunne & Raby realized their ambition of inserting their work into actual domestic environments, to allow for critical reflection by a using, rather than merely a viewing, public. They fabricated eight pieces of furniture, using MDF in as pared-down a way as possible, "to get against the emphasis on form." [86] Each piece gave material shape to an aspect of the anxiety surrounding the presence of electromagnetic fields in the home, and could be used as a tool either to measure their presence or to protect users from them. "Compass Table" contained twenty-five magnetic compasses that twitched or spun when an electronic product such as a laptop computer was set upon it. "Nipple Chair" incorporated a sensor that caused two nipple-like protrusions in the chair's back to vibrate in the presence of electromagnetic fields, making the sitter feel as if the radio waves were entering his or her torso. "Loft" comprised a ladder topped with a box that was lined in lead to allow for the storage of sensitive magnetic recordings. "Electro-draught Excluder" was a foam-lined shield that provided only a false semblance of protection from electromagnetic radiation, but that users could place between themselves and a television or computer to create a "sort of a shadow—a comfort zone where you simply feel better." None of these objects actually removed or counteracted electromagnetic radiation, but they could, as placebo devices, the designers hypothesized, "provide psychological comfort and, as such, reinforce the role of design criticism as therapy." [87]

As one of five finalists of the newly inaugurated Perrier-Jouët Selfridges Design Prize, in the spring of 2001, Dunne & Raby were able to display their work in the windows of Selfridges department store in central London, exposing it to an estimated 1.7 million passersby. They used the opportunity to present their Placebo furniture/objects, in an ideologically problematic yet expedient conflation of commerce's co-option of design, and their work's intended critique of design's commercial focus. Using notices in the Selfridges windows and advertisements in a London listings magazine, Dunne & Raby solicited individuals to adopt one of the "Placebo" objects and live with it for several weeks. Once their allotted time with the object was up, Dunne, Raby, and the photographer Jason Evans visited their homes to interview them about their experience of living with the object, and to photograph them interacting with it.

In this example of critical design in action, the design criticism occurred at more than one juncture. In themselves, the ambiguous Placebo objects—hybrids of furniture and appliances—provoked questions about their use. The criticism also occurred during the use of these objects. The people who lived with them experimented with putting them in different places in their homes, and reflected on the presence of invisible electromagnetic fields brought to their awareness through the physical form of the objects. Next, the objects' potential for constituting criticism was made available to others, through the extensive documentation of the project—stylized photographs of the adopters interacting

4.16
"Compass Table," in domestic setting, part of
"Placebo" project by Dunne & Raby, 2001,
photographed by Jason Evans. Courtesy of
Dunne & Raby.

4.17
"Draught Excluder," in domestic setting, part
of "Placebo" project by Dunne & Raby,
2001, photographed by Jason Evans. Courtesy
of Dunne & Raby.

PLEASE TOUCH THE CRITICISM

with the objects, and interviews which elicited the questions the objects had raised for them. These photos and interviews were published in the book *Design Noir*, and exhibited in multiple exhibitions around the world.

CONCLUSION

Dunne & Raby's attempt to break free from the gallery had been short-circuited, and in fact their association with the exhibition as a format only intensified. Their ideas on critical design were dispersed through a profusion of exhibitions throughout the 2000s devoted to the topic, including, in 2007, "Don't Panic: Emergent Critical Design" at London's Architecture Foundation, and "Designing Critical Design" at Z33 in Hasselt, Belgium. Their teaching, and later directorship of the RCA MA course in Computer Related Design (which was later renamed Design Interactions), also spread their ideas, with students such as James Auger, Noam Toran, and Elio Caccavale embarking on their own explorations into critical design.

Dunne & Raby were nomads operating between the conventional spaces of discourse in the 1990s. Through their connections to Japan, their critique of British manufacturing, and the global diaspora of their students and exhibitions, they did not belong to the conception of a creative British national identity being espoused by New Labour. As academics, they countered traditional notions of what constituted research; as practicing designers, their work fell outside the driving concern of commerce. Dunne & Raby's work seemed to fit well into the context of art practice, yet they rejected (or at least, attempted to reject) the art gallery as a site. Moreover, as critics, they existed outside the conventions of the design media.

By using the space of an exhibition venue, or the contours of a designed product, itself as a means of questioning social norms and the demands of industry, curators and designers in the late 1990s contributed to an ongoing destabilization of design criticism. As the design exhibition eschewed text, relied more on atmospheric impressionism, and became quieter, the product hybrids of the critical design genre became more vocal, literary and poetic. At a time when the design press appeared to have been subsumed by lifestyle marketing, the exhibition and the designed product became conduits for, and embodiments of, criticism. The advent of the design blog, discussed in chapter 5, would continue to open up the sanctum of design criticism and draw into question still further the role of the professional critic.

Customer Review

1,267 of 1,307 people found the following review helpful:

★★☆☆☆ **An expensive way to smell poo**, September 18, 2007

By <u>N A "Cat Lover"</u> ☑

Cat Genie takes the small unpleasantness of daily cleaning the litter and it saves it up and releases that unpleasantness as one big unscheduled, unpleasant inconvenience every week or two. Advanced monitors will ensure that the device failure will occur during the workday, as you prepare for your important meeting with your prospective client. Nothing like cleaning out wet cat poo in your nicest suit. Or, you may be pleasantly awoken in the middle of the night by the repeating three beeps of "there's poo and hair in the hopper." You will become more familiar with your cat's feces every day as the cat genie gently fills your home with the aroma of baking excrement. Plus, you get to pay over $300 for technology that was "designed" and built for less than $2. The "processor" unit was designed in 1967 and allows all the functionality of the most advanced microchip devices of its era. It has both on and off modes. (Note: off mode available only while unplugged.)

Actually, the real reason for the high cost of the device is to cover the costs of all the customer support that they must provide and to cover the costs of all of the returned units. The question is not IF, but WHEN you will find yourself hunched over your cat's feces floating in a pool of fetid water, picking small plastic pellets out of the opaque, pungent water with your fingers so that you can get the device put back together.

And your cats will thank you by depositing their love bundles beside the machine that's half filled with water and beeping away forlornly if you happen to be away when it fails.

We have three cats, they had no trouble adjusting to the machine over about a week. The small plastic pellets getting everywhere in the house is not really any big deal. Roomba takes care of most of them well. We've now had the machine for three months. We received a replacement base last week for a leaky drain hose. We've called their customer service line enough times that we now know the "secret" diagnostic techniques of their experts. We don't know if we're going to keep it or return it. If we keep it, we're definitely going to install an exhaust fan in the laundry room, and set it to a timer to go when the unit is on. For some reason there are little bits of poo that fall between the tines of the hopper, and they get slow baked every time the unit dries itself. The stench is really outstanding. It's hard to describe. I'm a doctor, and I've rarely ever smelled anything so bad.

My recommendation is to wait for the next generation cat sanitation solution. That device will need to be a complete redesign to solve the myriad of problems with this unfortunate device. To say something positive, the customer support line is manned by kind, well-meaning kids who really do feel badly that you're having a hard time with your mechanical poo soup maker.

Review Details

Item

CatGenie-Self Washing, Self Flushing Cat Box
★★★☆☆ (255 customer reviews)

5 star:	(102)
4 star:	(44)
3 star:	(23)
2 star:	(24)
1 star:	(62)

Out of stock

[Add to Wish List]

Reviewer

N A "Cat Lover"

Location: Tampa, FL

New Reviewer Rank: 75,466
Classic Reviewer Rank: 530,510
See all my reviews

5.1
Amazon customer review of CatGenie Self-Washing, Self-Flushing Cat Box, by NA Cat Lover, Amazon.com, September 18, 2007.

5.2
CatGenie Self-Washing, Self-Flushing Cat Box.

5

The Death of the Editor: Design Criticism
Goes Open Source, 2003–2007

In May 2007, NA "Cat Lover," from Tampa, Florida, bought a self-washing, self-flushing cat toilet. Moved to share his experience of the product with other potential customers on Amazon, his review began like this: "Cat Genie takes the small unpleasantness of daily cleaning the litter and it saves it up and releases that unpleasantness as one big unscheduled, unpleasant inconvenience every week or two. You may be pleasantly awoken in the middle of the night by the repeating three beeps of 'there's poo and hair in the hopper.' You will become more familiar with your cat's feces every day as the cat genie gently fills your home with the aroma of baking excrement."[1]

In the mid-2000s, Amazon.com extended its customer reviewing options to products as well as books. The site allowed consumers to review and rate their purchases using a five-star system. According to Amazon, there were a mere three suggestions to bear in mind when writing a "good review": "Be detailed and specific; What would you have wanted to know before you purchased the product? Not too short and not too long. Aim for between 75 and 300 words." By 2007, within these stark parameters a colorful genre of product criticism began to flourish, using many of the tools in the design critic's repertoire.

At 534 words, NA "Cat Lover"'s full review flagrantly disregarded Amazon's suggested word limit. It is detailed and specific, however. And it does provide technological and historically informed commentary, albeit tongue-in-cheek: "Plus, you get to pay over $300 for technology that was 'designed' and built for less than $2. The 'processor' unit was designed in 1967 and allows all the functionality of the most advanced microchip devices of its era. It has both on and off modes. (Note: off mode available only while unplugged)."[2]

The reviews that went beyond the standard fare on Amazon in the late 2000s were those that evoked worlds in which absurd products, like a self-flushing cat toilet, made sense. The author Geoff Dyer has given the label "imaginative criticism" to the mode in which he wrote a collection of semi-fictional riffs on the lives and works of jazz musicians. Instead of merely describing saxophonist Lester Young's "wispy, skating-on-air" tone, for example, Dyer paints a picture of everything that he imagines having led up to that tone: the untouched plates of Chinese food in Young's hotel room, the non-ringing phone, the gins with sherry chasers, his porkpie hat and cologne bottles on the bedside cabinet, and the condensation on the hotel window as he gazes across Broadway at Birdland. In the preface to his book *But Beautiful*, he writes:

> Before long I found I had moved away from anything like conventional criticism. The metaphors and similes on which I relied to evoke what I thought was happening in the music came to seem increasingly inadequate. Moreover, since even the briefest simile introduces a hint of the fictive, it wasn't long before these metaphors were expanding themselves into episodes and scenes. As I invented dialogue and action, so what was emerging came more and more to resemble fiction. At the same time, though, these scenes were still intended as commentary on either a piece of music or on the particular qualities of a musician.[3]

Another Amazon product review that exemplifies the principles of imaginative criticism, as described by Dyer, is a rather brilliantly sarcastic testimonial of a T-shirt adorned with three airbrushed wolves howling at a spectral moon. The reviewer, B. Govern, exploits our familiarity with the traits of the demographic he supposes would be likely to wear such a T-shirt: "This item has wolves on it which makes it intrinsically sweet and worth 5 stars by itself, but once I tried it on, that's when the magic happened. After checking to ensure that the shirt would properly cover my girth, I walked from my trailer to Wal-mart with the shirt on and was immediately approached by women. The women knew from the wolves on my shirt that I, like a wolf, am a mysterious loner who knows how to 'howl at the moon' from time to time (if you catch my drift!)."[4]

Even though it functioned as a bustling bazaar, visually the Amazon site was devoid of any images of people, or the circumstances in which the goods might be used. These user reviews provided the disembodied objects with human context—verbal *mises en scène*—in which they could be imagined more vividly. They used satire and narrative to entice the reader into a very particular world and then, by providing enough convincing detail, they persuaded the reader to stay. The detail reassured the reader that the authors had actually used and reflected on the product in question—that they actually cared. In the case of the CatGenie, it was the way NA "Cat Lover" noted the three beeps of the machine's alarm; with the T-shirt, it was the accumulation of brands and entities that accessorize the T-shirt wearer's lifestyle—Mountain Dew, tube socks, Wal-mart, crystal meth, and the courtesy scooter—in a neat subversion of the status symbols deployed in Jay McInerney and Bret Easton Ellis novels, such as naturally sparkling mineral water, Mont Blanc pens, Barney's, cocaine, and BMW 320is.

Amazon product review software also included a function that allowed a user to rank other people's reviews. Ranking was determined only with the criterion of "helpfulness," a quality that was clearly selected for the way in which it impelled one to practical action—and specifically, the act of consumption—rather than contemplation. The term "helpfulness" can be seen as a descendant of a modernist, instrumentalist stock of vocabulary used to evaluate design, which included such terms as "utility," "function," and "purpose," and had, over the course of the twentieth century, seeped over into the evaluation of design criticism itself.

The readers of Amazon product reviews of the late 2000s were also invited to comment on them and to add their own reviews in response, creating a kind of self-conscious metadiscourse around the practice of online reviewing. The Three Wolf Moon T-shirt review garnered thousands of responses and new reviews that emulated the style of the original. It also scored its author Brian Govern, then a Law student at Rutgers, national news coverage, a dedicated song and video, and many thousands of what were called "diggs," the currency of a user-driven social content website with widgets available for blogs and websites, founded in 2004, which allowed people to vote web content up or down. Registering for the service allowed people to become part "of the editorial process by digging and burying stories," and a "digital media democracy," the site averred in February 2007.[5]

Design critics have frequently complained of the experience of writing into a vacuum and having no idea if their point hit home or not. With the reflexive online discourse which emerged in the mid-2000s, some reviewers became minor celebrities as they rose in the reviewer rank through a voting system, and it was possible to analyze the impact of a review on several levels: how it affected sales of the product and related products, how much it was read and responded to, and how many people were moved to write their own reviews. Edgar of Baltimore wrote a narrative poem in the style of Edgar Allan Poe's "The Raven" about the product listed on Amazon as Tuscan Whole Milk 1 gallon 128 fl oz.[6] B. Smith, remarking on the reviews spawned by it, says: "After reading a few hundred, I had to compose my own. I still check the site for new reviews."[7]

Online reviewing was not limited to the written form. With a simple video camera and a tripod, a self-styled "product guru" like Mike Mozart was able to review hundreds of toys and products which got millions of views on YouTube and thousands of responses in the form of comments and other videos. His video review of a phallic pink Dora the Explorer AquaPet toy, which used exaggerated incredulity as a key rhetorical tactic in exposing its folly, had, by 2008, been watched more than a million times, and generated 15,000 comments and further videos.[8]

In some ways, the work of NA "Cat Lover," Brian Govern, Mike Mozart, and their ilk represents not a deviation from the true enterprise of design criticism but, rather, a logical extension of a democratizing impulse that, for its pioneers, has always been at its core. Ever since the early 1950s, when design criticism emerged as a genre in its own right alongside the industrial design profession, design critics have said that one of their main goals is to enable their readers to perform their own criticism.

In fact, in at least one case, it worked out exactly like that. In 1958, Judith Ransom Miller, an *Industrial Design* magazine reader and mother of four boys with large feet, sent in a manuscript of an article about the experience of being a consumer of socks via the Sears Roebuck Catalog. Ralph Caplan, who was editor at the time, published the article and later hired her as the magazine's West Coast correspondent, in which role she contributed further reviews, of the International Design Conference at Aspen, in particular. In framing her socks article in the magazine, Caplan observed:

> Here was a consumer who had something to say to designers, and could say it. Even the remoteness of the subject seemed a point in its favor, for the sock, as its author points out, is a "pure" item. In the belief that consumers should be heard as well as sold, I.D. dispatched a letter saying, "OK, you win. Who are you?" The answer: "I am a catalog consumer with a clinical turn of mind, interested in catalog merchandising as a means of modifying the design of some industrially produced goods, and as a vehicle for influencing the quality of consumership."[9]

Reyner Banham, writing in general-interest magazines such as *New Society* and *New Statesman*, made his subject matter accessible and his critical process visible, with a view to empowering the casual observer to comment on their own designed environment. In a 1983 essay, "O Bright Star," about the design of a sheriff's badge, for example, he described his research and evaluative process step by step, from the moment the decoration of the badge excited his curiosity, and he identified "the problem of who designs sheriffs' stars," through his dogged tracking of the source of its design and manufacture via libraries, police authorities, a factory's pattern shop; and finally to his realization, as the result of an overheard telephone call, that the badge was, in effect, designed by the Acme Star and Badge Co. secretary.[10]

Another writer discussed in this book, who was committed to the transparency of the critical process and the democratization of design criticism, was Jane Thompson, who coedited *Industrial Design* magazine in the late 1950s and pragmatically analyzed washing machines, cutlery, and chairs, drawing as much from her experience as a user of these products as from her connoisseurial training. Thompson reflected that she believed her critical writing of the period was about "trying to explain something so that the other person can have an opinion or evaluate it as well as you."[11] This pervasive desire to open up the mechanisms of design criticism contributed to its precarious and contentious position on the recreational–professional continuum of online media.

Design criticism is, by necessity, more self-aware of its proximity to the marketplace, its complicity with commerce and consumerism, than are other critical genres like art or literature. The incipient strain of amateur design criticism, located at the heart of the biggest online marketplace, illuminates and typifies many of the issues that were central to the reshaping of criticism's status and identity in the early twenty-first century. They included the differences between review and critique, recreation and professionalism, populism and elitism, production and consumption, as well as the role of ethics, consumerism, the nature of work, and time.

INTRODUCTION

By the mid-2000s, design criticism had found new means for dissemination in online forums such as blogs and websites. Seen by many as a democratizing force, helping to open up the previously inaccessible realms of criticism, these forums elicited new forms of writing and engagement, and brought into sharp focus many of the concerns that had been gathering during the previous decades over the purpose, quality, and format of design criticism, and its ability to connect to its public. Printed magazines used the art direction of page layout and sequencing and editorial devices such as hierarchy, juxta-position, highlighting and framing to set the stage for debate on their own terms and in their own timeframe. By contrast, the new aggregated context for reading in the era of social networking—or Web 2.0, as it was termed—had no temporal or spatial limitations; it grew and spread rhizomatically even as you read it. A piece of criticism might begin with a short provocative salvo in the main post of a blog and continue via a back-and-forth exchange in the Comments section. It could then migrate via links to other sites, its concerns highlighted in tagclouds and RSS feeds. This reading experience challenged the longstanding authority of editors and authors, and conferred new responsibilities on readers and commenters who, with a click of the "publish" button, could also become authors. Design criticism became increasingly fragmented, with multiple micro-constitu-encies, rather than recognized publishers or institutions, hosting and feeding the multiple conversations.

The perceived virtue of an open-source media landscape was that anyone with an opinion and an audience might contribute to critical discussion through their own blogs or those of others. Rebecca Blood, for example, like many other blog evangelists of the period, believed "in the power of weblogs to transform both writers and readers from 'audience' to 'public' and from 'consumer' to 'creator.'"[12] In her 2006 essay "Blogs: The New Public Forum," Sabine Himmelsbach concluded that the large numbers of bloggers indicated a metamorphosis of passive consuming readership into an actively participating public. She used optimistic terms and phrases such as "utopia," "global forum," "electronic agora," "democratic instrument," and "the possibility of exerting direct influence," which were typical of those in circulation at the time.[13]

Among the commentators who were less enthusiastic about the effects of blogging culture was *Boston Globe* columnist Maggie Jackson, whose book *Distracted: The Erosion of Attention and the Coming Dark Age* warned that modern society's inability to focus, manifested in blogging by its short-form writing and the multiple opportunities to link away from the original text, heralded an impending Dark Age. She believed that "Amid the glittering promise of our new technologies and the wondrous potential of our sci-entific gains, we are nurturing a culture of social diffusion, intellectual fragmentation, sensory detachment."[14] Digital media entrepreneur and journalist Andrew Keen was con-cerned about the "consequences of a flattening of culture that is blurring the lines between traditional audience and author, creator and consumer, expert and amateur."[15] Keen saw the mass amateurism of society as an insidious threat to both culture and the economy. Mainstream professionally produced media, he suggested, "provides us with common frames of reference, a common conversation and common values." In a filter- and editor-free, individualistic Web 2.0 world, however, long-held social values, such as the belief in a common culture, became fractured and were perceived as irrelevant:

> Wittingly or not, we seek out the information that mirrors back our own biases and opinions and conforms with our distorted versions of reality. We lose that common conversation or informed debate over our mutually agreed-upon facts. Rather, we perpetuate one another's biases. The common community is increasingly shattering into three hundred million narrow, personalized points of view. Many of us have strong opinions, yet most of us are profoundly uninformed.[16]

In the mid-2000s, design publishing was foundering due to reduced advertising revenue and dwindling readership for print magazines. *House & Garden*, *Domino*, *Step*, *Grafik*, *Emigre*, and the *AIGA Journal* ceased publication, followed by *I.D. Magazine* in January 2009. Others attempted to forge online presences, with the British visual communication magazine *Creative Review*'s efforts among the most successful. In the increasing void left by such design magazine's closures, new websites devoted to design sprang up. Without the historical baggage, nor the overheads of traditional publishing companies, sites like Dezeen, Core77, and DesignBoom emerged as news and information sources for the design community. Largely image-based, celebratory in tone, and heavily dependent on a flow of promotional press releases to feed their fast-paced content streams (Core77 posted about 15 articles a day in 2007 and Dezeen posted around 10 per day), such sites were driven by the competitive need to be the first to identify new products and ideas, and to marshal traffic for their advertisers.

The design community was quick to adopt blogging as a medium. Before default blogging software became readily available, the blog pioneers wrote their own code, and thus tended to come from the worlds of technology and design—mostly website production. Individual designers, especially those interested in web culture, such as Jason Kottke and Khoi Vinh, began blogs through which they commented on technological developments and gained large readerships. Vinh, for example, began blogging on his site Subtraction.com in 2000, using Blogger as a publishing tool. In 2003 he switched to Movable Type and began writing longer daily posts on technological developments germane to digital design, his own design practice, and design issues more generally, attracting a daily readership in the tens of thousands and some advertising revenue.

Blogging about product design had become a profitable enterprise, because reviews led readers to make purchases. As political scientist Jodi Dean put it, "by starting their own blogs, hiring bloggers, and participating in discussions related to their products, companies could market in another mode."[17] Product design firms identified prolific bloggers such as Grace Bonney at Design Sponge or Tina Eisenberg at Swiss Miss, and sent them samples to review. By 2009, 70 percent of bloggers said they blogged about brands.[18] The sense that one's work as a blogger could be monetized accelerated what Dean calls "blogging's centripetal momentum."[19] Writing about the culture of design or about graphic design, in which there was no identifiable product to purchase, however, did not lend itself to such an economic model. Most graphic design blogs of the mid-2000s still operated in the gift economy. Bloggers exchanged reciprocal links, and helped to promote each other's blogs. Some design blogs, especially those derived from magazines such as *Metropolis* or from commercial concerns such as Mediabistro, sold advertising or posted sponsored content. Some individual bloggers, such as John Thackara and Joe Clark, attempted to initiate micro-patronage by which readers would donate contributions to support their work. Some blogs were funded by grants and endowments. Mostly, however, blogging about graphic design was accepted as an unpaid hobby, where the absence of the framing role of an editor and the lack of training or experience in critical writing were considered to be liberating.

Speaking up and talking back: quality control in the era of the design blog

Designers with strong opinions found a new and welcoming home in the blog Speak Up, which was founded in 2002 by the graphic designers Armin Vit and Bryony Gomez-Palacio, then in their early twenties and based in Chicago, where Vit worked at a small agency called Norman Design and Gomez-Palacio at Bagby and Company. In the evenings they blogged about design.

They adopted a plain-speaking, approachable voice for their blog, recasting conventional website navigation headings in conversational terms: "So what exactly is this place?" (for the more usual "About" section) and "Let me go please" (instead of the more prosaic "Unsubscribe"). To introduce himself, Vit wrote: "And what makes me a design critic? Nothing really. I just need an outlet to speak up, and hopefully somebody will listen and would like to say something too."[20]

Vit's posts were short, opinion-based observations, articulated in off-the-cuff rushes, mostly about new design work, the activities of professional organizations such as the American Institute of Graphic Arts, or visual tropes he had noticed. Encouraged by the enthusiastic response of the design community—some posts gathered more than 200 comments—Vit further emphasized the "open dialogue" aspect of the site and started to invite others to contribute to the blog, including designers Jason A. Tselentis, Marian Bantjes, Tan Le, Graham Wood, and Mark Kingsley. By the time Vit and Gomez-Palacio moved to New York in 2005, Speak Up was an active online community generating multiple posts per day, and many hundreds of comments.

By May 2007 things had quieted down, and most of the pieces posted that month garnered only a smattering of comments. They dealt with topics such as a new contest-structured website that connected companies with video makers; design workshops as a genre; and a favorable review of Steven Heller and Mirco Ilic's book *The Anatomy of Design*. There was one piece, however, posted on May 4, 2007 by New York-based designer and creative director Mark Kingsley, that generated 125 comments and refocused the design blogging community's attention back onto the blog as a medium.[21]

Mark Kingsley grew up near Buffalo in upstate New York, and studied graphic design in the mid-1980s at Rochester Institute of Technology, where he received a broad-based education in visual culture. In the early 2000s, as Speak Up was gaining momentum, Kingsley and his wife ran a small boutique design firm based in Chelsea, specializing in music packaging and branding for cultural organizations such as Summer Stage. Around 2005, Vit invited him to become a Speak Up author, which involved contributing at least one post a month and being an active presence among the commenters.

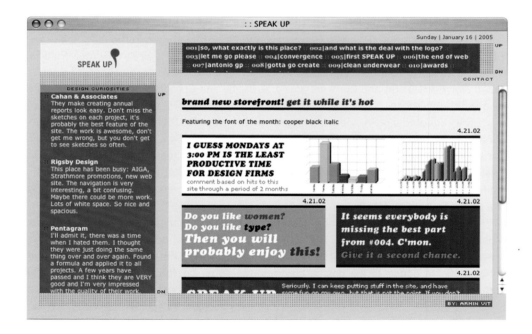

5.3
Speak Up blog homepage, by Armin Vit and
Bryony Gomez-Palacio, January 16, 2005.
Courtesy of Armin Vit and Bryony Gomez-Palacio.

THE DEATH OF THE EDITOR

On a spring morning in 2007, Kingsley received an email from Vit, who was on paternity leave from Pentagram at the time, asking him to take a look at a column in the May/June issue of *Print* magazine written by the British critic Rick Poynor. Could he respond on behalf of Speak Up? Kingsley bought the issue, took it to the Empire Diner on the corner of 22nd Street and 10th Avenue, which he liked to frequent since he learned that Einstein had once eaten there, and sat with a large cup of coffee, reading Poynor's column. Leaving his coffee half-finished, Kingsley leapt up from his corner booth and strode back to his studio, already in his head drafting the impassioned response he would post on Speak Up. Kingsley had not contributed for a while, but Poynor's provocation was enough to bring him back onto the online soapbox.

Easy writing

By 2007, aged fifty, Poynor had authored twelve books, edited and contributed to many more, and published three volumes of collected essays about design. In the late 1980s he had been deputy editor of *Blueprint*, and in 1990 he became founding editor of *Eye*, an international quarterly journal on graphic design, and probably the best-respected publication on the topic at the time. In 1997 he gave up the editorship and continued to write as a freelance critic from his home in a South London suburb. In 2003 Poynor joined Connecticut-based designers and publishers Jessica Helfand and William Drenttel and New York Pentagram partner Michael Bierut to become a founding member of the online forum Design Observer. He left the site in 2005, frustrated with his co-founders' resistance to editing the contributions, the lack of remuneration for writers, and the unmediated nature of the intellectually shallow and off-topic comments. He continued as contributing editor to several publications in both the US and the UK, including *I.D. Magazine* (as *Industrial Design* magazine was renamed in the 1980s), *AIGA Journal of Graphic Design*, *Graphis*, *Eye*, and *Blueprint*, but in 2007 his primary column was in *Print* magazine, where he had been a contributing editor since 2000.

Poynor's 1,300-word column, titled "Easy Writer," published in the May/June issue of *Print*, argued that without editors to help shape their articles, design bloggers were unable to produce writing of the same standard as print publications.[22] After five years of operation, Poynor wrote, Speak Up had failed to produce any high-quality design criticism—a type of writing he believed to be characterized by its "range of commentary, depth of research, quality of thought," among other attributes. He used the recently published edition of graphic design writing anthology *Looking Closer 5* as his litmus test: in a collection of forty-four examples of supposedly exemplary writing published in the past five years, only four derived from blogs, and none of those was from Speak Up—even though, as Poynor observed, "according to Vit, Speak Up alone has produced more than 1,500 posts." The period of time the anthology covered, 2002–2007, was the same period that Speak Up had been in existence. "It has been quite common during this time to suggest that blogs represent the great hope for a thriving new critical debate, a place where an ambitious upcoming generation of design writers can sharpen their critical skills and prose. I have made the same claim, or at least expressed the same hope, a few times myself," Poynor wrote, summarizing the widely felt optimism that still surrounded blogs at this time.[23]

What was to blame for what he saw as the blogs' poor performance? In Poynor's view, the biggest single problem with blogs was the absence of editors. As both a writer and an experienced editor himself, he knew the kind of work that went on behind the scenes to produce something "fit for print." The editorial effort, for Poynor, was about dealing with "larger issues of content and the development of a strong argument," as well as "with details of copywriting."[24] Most Speak Up contributors had never worked with an editor before, had never benefited from being forced to answer difficult questions, research, rewrite, polish, and fact-check a piece. Editing, in the sense of giving shape to the publication as a cohesive entity, was not a factor for consideration in the online environment. Poynor, however, saw this approach to the production of writing as both amateurish and unexpected, since "designers are quick to reject amateurishness within design; exactly the same considerations should apply to editing and writing."

The problem of the lack of editing, in Poynor's view, was compounded by the lack of remuneration for writers in the online sphere—the hobbyist nature of their enterprise: "Research will always suffer where there is no cash to fund it."[25] He believed that without payment to offer, it is hard to sustain contributors' involvement over time, or to attract established writers, who depend on fees to make a living. To Poynor, the fact that design writing of the early twenty-first century was not valued monetarily was among the biggest threats to its survival.

Hopes for Speak Up and other blogs' output had been set too high, Poynor opined. Poynor took Vit to task for making "grandiose claims about how critical Speak Up had been." Speak Up may have generated "sharp and revealing exchanges," he wrote, but printed publications, he argued, ultimately provided a better environment for good writing and good criticism to flourish.[26] Poynor was acutely self-aware of his role as a critic and wrote several articles about what he saw as the poor health, and even the "death," of design criticism in the sense of an authoritative, ideological cultural endeavor which, through an unmasking and analysis of design's social influence, "forces open people's eyes," and where the writing itself "offers pleasures of its own."[27]

Tourette's syndrome and bar brawls

Despite his lack of experience as a writer, Mark Kingsley took pride in his writing for Speak Up, choosing to work in the early hours of the morning, "when one's defenses are down," in long, "Tourette's-like" streams of "automatic writing" that he compared to the creative process of the American composer Robert Ashley.[28] He considered his writing to be

THE DEATH OF THE EDITOR

visual, and often used images, as links or inserted into the text, as integral components of his argument, a mode that was well suited to the blog medium. Kingsley studied French, and liked to use foreign words and Latin or obscure terms to give his readers pause, and often used etymology to give ballast to a point (a tendency that he later dismissed as a "crutch").[29]

Kingsley's fiery 1,400-word missive was posted on May 4, 2007, with the title "Rick Poynor: Ipse Dixit."[30] The Latin phrase means, literally, "he, himself, said it," which in the field of logic refers to an unproven assertion that the speaker claims is true based only on his or her authority. Kingsley objected to the causal connection Poynor had made between the lack of Speak Up essays in the *Looking Closer* anthology and the quality of Speak Up essays. While he admitted that Speak Up may have lacked professionalism, he believed that its authors' passionate writing, their ability to engage their audience, and the way in which novice commenters were being educated through participation, outweighed any of its deficiencies. "It's a mess, there's a lot of shitty prose to wade through, and many of the ideas are half-baked," Kingsley wrote. "But at its best, Speak Up makes that emotional connection."[31]

Kingsley did not probe Poynor's main assertion—that the site suffered from its lack of editing—and focused instead on what he saw as Poynor's blatant miscomprehension of the qualities of a blog, which Kingsley considered to be "a unique aesthetic—not quite conversation, not quite a measured exchange of belles lettres," but something else not easily compared to a printed magazine.[32] Kingsley felt that "Rick's impulse was misdirected. He didn't understand how blogs worked and that the important thing about Speak Up was its intent, its intensity, and the fact that, like after a good jazz session, you could see blood on the floor."[33] Kingsley wrote to stimulate the views and objections, new arguments, and references of his peers. He relished the presence of his audience, which he considered more akin to a theater audience than a readership, referring to them in one comment as the "peanut gallery." He always engaged in the comments, entering into long sparring matches with whoever was willing to take him on, enjoying the performative aspect and public nature of such exchanges. A blog thread would usually run its course in twenty-four to forty-eight hours, further emphasizing its resemblance to a live performance. In his view, in a blog "the comments *are* the editing. It's more of a discussion than an ex cathedra missive. Shouldn't you be present for the comments and allow them to try to change your mind and to try to change theirs? It's through that conflict where things are built."[34] As the comments started to pour in, in response to Kingsley's post, they echoed his belief in Speak Up's unique ability to make an emotional connection to its readers. The commenters frequently compared the "detached" style of Design Observer writers with the "passionate" approach of Speak Up. The erudite pieces on Design Observer were meant to be admired from a distance, Speak Up commenters averred, while at Speak Up you could plunge in and take part in the conversation, no matter how little knowledge you had on the topic at hand.

5.4

"Rick Poynor: Ipse Dixit," by Mark Kingsley,
posted on Speak Up on May 4, 2007, in response
to Rick Poynor's *Print* column. Courtesy of
Armin Vit.

The very first comment, from "ben," asked: "Is Rick saying Design Observer is better than Speak Up?" Kingsley responded by saying: "Ben, it is inferred," and then picked out the quote from Poynor's piece most likely to incite the Speak Up community of commenters: "but the main point is he has found what we do here to be unworthy. Readers' comments are described as having to 'wade through a lot of bilge to fish out sharp and revealing exchanges.'"[35] Subsequent commenters made the same assumption as "ben," and added their own views on the comparison between Speak Up and Design Observer. It was obvious that very few of them had read Poynor's original essay—which did not compare Speak Up with Design Observer—but depended instead on Kingsley's summary of it. Derrick Schultz characterized Speak Up as a "learning environment" for emerging writers like himself. "Sometimes the articles on here are eye-rolling from an audience standpoint, but they are eye-opening for the writer," Schultz wrote, revealing another of Poynor's charges against unedited blogs—the danger of succumbing to "self-indulgence."

One of the more thoughtful comments was posted by the designer Marian Bantjes, a regular Speak Up author. It exposed what she saw as a flaw in Poynor's original article more precisely than Kingsley had done in his post: "Speak Up is *not* an online magazine or a journal, but a place where people gather: much closer in analogy to a bar than a publication." The purpose of Speak Up, in Bantjes's opinion, was not to create "perfect" articles along the lines that she thought Poynor had described in *Print*, but rather to engage commentary. Her point was that, while Poynor saw the absence of Speak Up essays in the *Looking Closer* collection as an indicator of the poor health of blog writing, Speak Up authors would never have expected their work to be found in such a publication. Vit echoed Bantjes's point in his own comment: "I couldn't care any more about not having anything in *LC5* than I do about missing an episode of *Dancing with the Stars*. Speak Up is a blog, and its place is the internet."[36] Speak Up posts—typically short, conversational bursts of commentary and opinion built around links—were not conceived as essays, and their success was not measured by the standards that Poynor used, but rather by their ability to hold the attention of readers who were usually at work, and to stimulate the most discussion. Blog writing has been described as "voicey," referring to its opinionated nature but also, more literally, to the blogs' goal, which was to reproduce the authenticity of direct, colloquial speech. As Jodi Dean has observed, "blogs, and even more so Twitter, catch oral communication in linear writing. Like a phone call or a text message, a post attempts connection. … Instead of judging blog posts as a literary form, it is more useful to consider them as a form of expression in between orality and literacy."[37]

Responding to a response

By May 11, Kingsley's article had generated more than sixty comments. At this point Poynor stepped in again, but instead of adding a comment to the Speak Up site, and with what was perceived as another slap in the face to the bloggers, he used *Print* magazine's website to publish his reply. "Telling that Poynor does not participate in the discussion where it happens, but elsewhere," remarked "ps" of this move.[38] In his response, Poynor methodically unstitched Kingsley's "false opposition" between "dull professional perfectionism and thrillingly passionate amateurism ... this is clearly nonsense—you can be both passionate and totally wrong-headed ... while passion does indeed help make an emotional connection with the reader, it's hardly the bedrock of good writing."[39]

Print's editor, Joyce Rutter Kaye, posted in the Speak Up comment thread a link to Poynor's response, and another geyser of comments erupted. Marian Bantjes drew attention again to Poynor's insistent focus on the initial posts rather than to the holistic experience of blog participation, which included—and for some began with—the comments thread: "To ignore the discussion as a huge part of the reading experience is to miss the point of a blog."[40]

Poynor has noted of his time at *Blueprint*, when he began to realize that the 1980s obsession with style was "almost always masking a hollowness," that he "was always very preoccupied with issues of worth and value." This belief in value was to do with his training in art history, but was also, he offered, "just probably what I'm like. It's the way I weigh up and measure things ... I think that's a pretty reasonable, indeed a standard, preoccupation for a critic to have."[41]

Meanwhile, Speak Up commenter Joe Natoli represented a view held by many Speak Up contributors: that professional writers were elitist, patronizing, and unnecessary; one could just as easily learn about design as well as how to write through writing comments, and not necessarily reading articles or even the original post. He added to the thread of comments generated by Poynor's response to Kingsley's post: "We don't need or want to be told how to think, we don't need these people to tell us what is of value and what isn't. Instead, we need to talk to each other, dive deep and learn from the exchange. All the responses above have done just that for me ... I've gained more valuable knowledge, inspiration and insight from Armin Vit and the folks who regularly post to Speak Up than I have from 10 years of reading the Poynors and Hellers of the world."[42]

They may have shared design as subject matter, but ultimately the Speak Up authors and Poynor appeared to be separated by an unbridgeable gulf. Poynor is a writer, respectful of the discursive and playful qualities of language, the craft involved in shaping an article, and the role of the editor in the process. Speak Up's approximation of a multi-vocal conversation provided a palpable example of the ways in which the Marxist poet, editor, and broadcaster Hans Magnus Enzensberger had predicted, in 1970, writing's demotion to "a secondary technique," and the ways in which electronic media "tend once more to make people speak."[43] It did so through its very title, the conversational and stream-of-consciousness writing style of its authors, the lack of concern for "good writing," the absence of editors, and the conscious attempt by its participants to approximate the atmosphere of a heated discussion at a bar.

THE DEATH OF THE EDITOR

Aftermath

Despite the talk of barroom brawls, in fact, even by May 2007, when the clash over Poynor's "Easy Writer" article took place, the trouble-making aspects of Speak Up had begun to recede; the views of its authors were moving toward those of the design orthodoxy. In February, Vit had written a repositioning post, suggesting that the blog was having a midlife crisis: "In the past twelve to sixteen months ... we've run out of questions and even perhaps out of steam. Some of us (authors) have gone from outsiders to insiders."[44] Vit was by then thirty, a member of Michael Bierut's team at Pentagram, involved in AIGA, and had recently become a father. Mark Kingsley freelanced for the Branding Integration Group at Ogilvy & Mather, and later went to work for Landor Associates as the global creative lead for the Citibank account. Even as it argued the virtues of its thriving community to Poynor, Speak Up was already nostalgic for its feisty youth; now, as Vit explained, the blog would turn its attention to "Design Relevance," a bland-sounding concept that echoed the language of contemporaneous press releases from AIGA, Cooper Hewitt National Design Museum, and other establishment institutions.

In April 2009, after 1,600 posts and 43,000 comments, Vit and Gomez-Palacio pulled the code on Speak Up, acknowledging that the blog was dying before their eyes. In a series of parting posts, Vit attributed their decision to a loss of sustaining power on the part of its founders and contributors, declining numbers of posts, comments, and visitors, and the fact that its energy had been diluted by its splinter blogs—Brand New, Quipsologies, and Word It. He reflected: "I also strongly believe that the kind of general-topic and long-form writing of Speak Up is just not as appealing as it used to be. With so many web sites devoted to quick bursts of visuals and the proliferation of short-message communication enhanced by Twitter and Facebook, it becomes increasingly hard to hold the attention of anyone."[45]

Meanwhile, *Print* magazine struggled financially as its readership dropped 50 percent to around 40,000. It underwent several changes of editorship in the late 2000s but ultimately survived, and Poynor continued to write his column. Vit became a regional juror for the magazine's design competitions, and Speak Up authors, including Debbie Millman, became regular contributors. In 2010, Poynor rejoined Design Observer as the author of his own blog, titled "Adventures in the Image World." By this time Design Observer was able to pay its regular writers through advertising revenue and a substantial grant from the Rockefeller Foundation. By being the sole author of this blog, albeit under the larger Design Observer Group umbrella, Poynor lacked the oversight and challenge of editorship he had told the Speak Up community was necessary to good writing. Nevertheless, he maintained a sense of mission that was in striking contradistinction to the premise of Vit's new brand-focused blog: "The more commerce attempts to corral and confine design and the image world for its own purposes, the more we need to seek out, savor and support work that connects with areas of experience other than lifestyle and celebrity—work that is awkward, offbeat, difficult, socially challenging, strange or fantastical and that offers vital, mind and spirit sustaining alternatives to the insidious, corporatized monoculture."[46]

The quality that made Speak Up resonate so powerfully with its readership of graphic designers—its amateurism—and the very attribute that Poynor had called out as being so problematic, was also what ultimately led to its demise. It appeared that an amateur means of production could not support sustained critical output, nor families, for that matter. Bloggers are human beings too.

CONCLUSION

Jodi Dean identified the summer of 2007 as the moment when the "bell tolled for blogging."[47] Even as the number of blogs steadily rose, and corporations became increasingly involved in blogging, "word spread rapidly that blogs had been killed by boredom, success, and even newer media." Dean saw in blogging's obituary a change in practice in which blogs became "elements of an inescapable circuit in which we are caught, compelled, driven."[48]

As the New York Times columnist Thomas Friedman has noted, 2007 was a notable year for the launch of the physical technologies of social media that further reinforced the "inescapable circuit" of transindividual communication. Apple launched the first iPhone, and Facebook (which had until then been confined to universities and high schools) and microblogging platforms Tumblr and Twitter gained mainstream traction. This combination of mobile internet connectivity and social networking services allowed users to express themselves faster and to reach larger audiences in the form of short updates, images, 140-character commentaries, or single-click republishing.

Friedman believes that even though 2007 may have been "one of the greatest technological inflection points in history," at the time its significance was obscured by the global financial crisis and recession of 2008 which prevented the instigation of the "rules, regulations, institutions and social tools people needed to get the most out of this technological acceleration and cushion the worst. ... A lot of people got dislocated in the process."[49]

Design criticism underwent its own process of dislocation in the late 2000s. As design magazines continued to fold, death rhetoric, similar to that being used about blogs, was used in connection to design criticism. Morbidly titled articles published on- and offline appeared with frequency, among them: "The Death of Graphic Design Criticism," "The Death and Life of Great Architecture Criticism," "Another Design Voice Falls Silent," "The Death of the Critic," and "Where Are the Design Critics?"[50] And yet, at the same time, new interest in design criticism was galvanized around the launch of programs, conferences, and online forums devoted to nurturing the discipline even, and perhaps especially, in its dematerialized state.[51] With the closure of so many design-specific publications, design criticism atomized, but it did not disappear; like blogging—the tone, authors, and posting speed of which was absorbed into the publishing ecosystem—it adapted to the new conditions of diaspora. Design was too amorphous and all-encompassing to be given its own section of a newspaper, to be ghettoized into professional and trade publications. As we have seen, design criticism came of age in the postwar period, driven by a need to boost national economies and bolster professional insecurities. Relieved of such responsibilities, and with a broader range of interconnected platforms at its disposal, design criticism could operate as a variety of cultural criticism, with design being not only the object of judgment, but also the point of departure for a critique aimed at society and the world.[52]

As Michael Rock observed in his 1995 discussion with Rick Poynor, "design criticism is everywhere, underpinning all institutional activity—design education, history, publishing and professional associations."[53] Following this logic, the list has only expanded in the intervening years to include the curation of exhibitions, the direction of conferences and events, the production of videos and podcasts—indeed, the intellectual choreographing of any kind of activity through which arguments about the successes, failures, meanings, and social and environmental implications of design might be articulated and debated.

Today, design criticism continues to be uncertain about how and where to gather its publics, and for what ends. Professional criticism's relevancy as a gatekeeper has been usurped by the irreversible realities of an instantaneous publishing landscape where, as Clay Shirky tells us, "everyone is a media outlet."[54] We trawl our daily streams, retrieving images, messages, tweets, videos, soundbites, links, and reconstructing these fragments into customized and personalized feeds, a tendency which MIT Media Lab cofounder Nicholas Negroponte had in 1995 prophetically dubbed the "Daily Me," and which results, troublingly for many of us, in our being exposed only to content we are already inclined to agree with.[55]

Now that our reading predilections are monitored so relentlessly, we signal our endorsement of certain pieces of design criticism, and the worldview they represent, not just with each comment, "share," and "like," but also with each page view, and even with our search terms. We aggregate our customized design criticism from the millions of users' perspectives on products, appliances, interiors, and services found on Amazon, Engadget, and Yelp, as well as from in-depth, reporting-based essays about designers and design ideas in general-interest publications like the *New Yorker* or the *Guardian*, from documentaries on YouTube, podcasts on 99% Invisible, from lectures, community meetings, protests, and from scholarly peer-reviewed papers on Academia.edu, for example. Despite the eclectic sources of this new reading-viewing-listening-participating experience, ultimately its distinctions are flattened. Our ability to discern the various textures of the authors' voices, the political and ethical worldviews of the commissioning agents, and the contexts of ongoing conversations and reference points—which were once all evident when criticism came in a publication-shaped package—has been traded in for the seemingly ideal conditions of instantaneous, pluralistic, democratically accessible, popular criticism currently being conducted in three billion interconnected personal microcultures.

Today, therefore, just like Dickens's dust mound denizens, and the many design critics that have followed them—Victorian design reformists, postwar good design anesthetists, throwaway economists, Drop City communitarianists, junk merchants and tastemakers of design-decade London, 1990s urban beachcombers, bloggers, Yelpers and American Pickers—it is we, the public ourselves, who must be the sifters, and not just of design, but also of what Brian Thill calls our "endlessly accumulated tabflab," the constantly updating, linking, tagging, and streaming piles of digital detritus.[56] Today, as we bear much of the responsibility to create and curate our own content, and as browsing culture is replaced by searching culture, we must be critical of the algorithms that seek to determine our preferences. On our screens, but also well beyond them in our streets, schools, studios, museums, and meetings, we must continue to make, identify, and use the kinds of design criticism with the potential to enrich the ways we think about design, to diagnose symptoms of harmful and wasteful practice, conduct informed salvage missions, and then illuminate paths to recovery.

NOTES

INTRODUCTION

1. Charles Dickens, *Our Mutual Friend*, <http://www.gutenberg.org/files/883/883-h/883-h.htm> (accessed October 20, 2016).

2. I. A. Richards, *The Principles of Literary Criticism*, ed. Paul Kegan (London: Trench, Trubner, 1924), vii. R. P. Blackmur, "A Critic's Job of Work," in *Selected Essays of R. P. Blackmur*, ed. Denis Donoghue (New York: Ecco Press, 1986), 19.

3. Zygmunt Bauman, *Wasted Lives: Modernity and Its Outcasts* (New York: Blackwell, 2004), 19.

4. Jonathan Crary, *Techniques of the Observer* (Cambridge, MA: MIT Press, 1993) 19.

5. Brian Thill, *Waste* (New York: Bloomsbury Academic, 2015), 3.

6. Ben Highmore, *The Design Culture Reader* (London: Routledge, 2008), 1.

7. Bruno Latour, *An Inquiry into Modes of Existence* (Cambridge, MA: Harvard University Press, 2013), 8.

8. Vilem Flusser, *Shape of Things: A Philosophy of Design* (London: Reaktion Books, 2012), 90.

9. Thill, *Waste*, 7.

10. Reinhold Leinfelder, "Assuming Responsibility for the Anthropocene: Challenges and Opportunities in Education," in "Anthropocene: Exploring the Future of the Age of Humans," ed. Helmuth Trischler, *RCC Perspectives* 2013, no. 3, 12.

11. Timothy Morton, *Hyperobjects: Philosophy and Ecology after the End of the World* (Minneapolis: University of Minnesota Press, 2013), Kindle ebook, 1.

12. Wolfgang F. E. Preiser, Harvey Z. Rabinowitz, and Edward T. White, *Post-Occupancy Evaluation* (Abingdon, Oxon: Routledge Revivals, 2015).

13. Susan Strasser, *Waste and Want: A Social History of Trash* (New York: Henry Holt, 1999), 3.

14. Reyner Banham, "Rubbish: It's as Easy as Falling off a Cusp," review of Michael Thompson's book *Rubbish Theory: The Creation and Destruction of Value, New Society*, August 2, 1979, 252.

15. Reyner Banham, "A Throw-away Esthetic," *Industrial Design*, March 1960, 65.

16. Clifford Humphrey, Speakers' Papers, IDCA 1970, International Design Conference in Aspen Records 1949–2006, 2007.M.7, Series 1, Box 28, Fol. 4–6, The Getty Research Institute, Los Angeles.

17. Claire Catterall, *Stealing Beauty: British Design Now* (London: ICA, 1999), 7.

18. Stephen Spender, "Thoughts on Design in Everyday Life," Design Oration of the Society of Industrial Artists, 1958, published in *SIA Journal*, January 1959.

19. Jürgen Habermas, "The Public Sphere: An Encyclopedia Article," *New German Critique 3* (1974), 49.

20. Jeffrey Meikle, "On the Ideal and the Real in Design History," *Journal of Design History* 11, no. 3 (1998), 194.

21. Deanya Lattimore, Diigo.com, June 26, 2008, <https://www.diigo.com/list/deanya/public-space-exhibitions> (accessed October 28, 2013).

22. Julia Kristeva, *Desire in Language: A Semiotic Approach to Literature and Art*, ed. Leon S. Roudiez, trans. Thomas Gora, Alice Jardine, and Leon S. Roudiez (New York: Columbia University Press, 1980), 91.

23. Beverly Long and Mary Susan Strine, "Reading Intertextually: Multiple Mediations and Critical Practice," *Quarterly Journal of Speech* 75 (1989), 468.

24. Ann Sobiech Munson, "Lewis Mumford's Lever House: Writing a House of Glass," in *Writing Design: Words and Objects*, ed. Grace Lees-Maffei (London: Berg, 2011), 119–132.

25. Peter Hall, "Academe and Design Writing: Changes in Design Criticism," *Design and Culture* 5, no. 1 (Spring 2013), 22.

26. Clive Dilnot, "The State of Design History, Part I: Mapping the Field," *Design Issues* 1, no. 1 (Spring 1984), 249.

27. M. C. Lemon, *The Discipline of History and the History of Thought* (London: Routledge, 1995), 109.

28. Richard Hoggart, "Schools of English and Contemporary Society," in *Speaking to Each Other* (London: Chatto & Windus, 1970), vol. 2, 256–257.

29. Terry Eagleton, *Literary Theory: An Introduction* (Oxford: Blackwell, 1983), 181.

30. John Simon, quoted in David Carr, "Some Highish Brows Furrow as a Car Critic Gets a Pulitzer," *New York Times*, April 8, 2004, <http://www.nytimes.com/ 2004/04/08/arts/08PULI.html> (accessed November 16, 2016).

31. Dave Hickey, *Air Guitar: Essays on Art and Democracy* (Los Angeles: Art Issues Press, 1997).

32. Reinhold Martin, "Critical of What? Toward a Utopian Realism," *Harvard Design Magazine*, no. 22 (Spring/Summer 2005), 5.

33. Andrew Wernick, *Promotional Culture: Advertising, Ideology and Symbolic Expression* (Thousand Oaks, CA: SAGE Publications, 1991), 195.

34. Roland Barthes, "The New Citroën," in *Mythologies* (London: Vintage, 1993), 88.

35. Ibid.

36. Ibid., 89.

37. Judith Williamson, "Inflated Intangibles," *Eye*, Spring 1999, 7.

38. Michael Horsham, "The Value of Confusion," in Catterall, *Stealing Beauty*, 13.

39. Walter Benjamin, *The Arcades Project*, N1, 9/458, quoted in Christopher Rollason, "The Passageways of Paris: Walter Benjamin's *Arcades Project* and Contemporary Cultural Debate in the West," <http://www.wbenjamin.org/passageways.html#fn76> (accessed October 20, 2013).

40. Walter Benjamin, "On the Concept of History," in *Gesammelte Schriften*, I:2 (Frankfurt am Main: Suhrkamp, 1974), trans. Dennis Redmond, at <http://members.efn.org/~dredmond/ Theses_on_History.PDF> (accessed October 20, 2013).

41. Lemon, *The Discipline of History and the History of Thought*, 71.

42. Ibid., 72.

43. Ibid., 112 and 43.

44. Giorgio Agamben, "What Is the Contemporary?," in *What Is an Apparatus? and Other Essays* (Stanford: Stanford University Press, 2009), 40.

45. Ibid.

46. Bruno Latour, "Has Critique Run Out of Steam? From Matters of Fact to Matters of Concern," *Critical Inquiry* 30 (Winter 2004), 246.

47. Bruno Latour and Peter Weibel, eds., *Making Things Public: Atmospheres of Democracy* (Cambridge, MA: MIT Press, 2005), 31.

48. Rick Poynor, "The Closed Shop of Design Academia," *Design Observer*, April 13, 2012; Matt Soar, "Rick Poynor on 'Design Academics': Having His Cake and Eating It Too," Matt Soar's blog, April 19, 2012; Hall, "Academe and Design Writing."

CHAPTER 1

1. Deborah Allen, personal interview, July 6, 2007.

2. Deborah Allen, "Cars '55," *Industrial Design*, February 1955, 82.

3. Richard Hamilton, "The Urbane Image," *Living Arts* (Institute of Contemporary Arts), no. 2 (1963).

4. Richard Hamilton, personal interview, February 23, 2007.

5. Richard Hamilton, *Richard Hamilton Paintings Etc. 1956–1964* (London: Hanover Gallery, 1964).

6. Ibid.

7. Reyner Banham, "Who Is This Pop?," *Motif*, Winter 1962/63, 3.

8. Ibid.

9. Reyner Banham, "A Throw-away Esthetic," *Industrial Design*, March 1960, 65 (originally published as "Industrial Design and Popular Art" in *Civiltà delle Macchine*, November 1955).

10. Reyner Banham, preface to *Design by Choice*, ed. Penny Sparke (New York: Rizzoli, 1981), 7.

11. Press release, June 22, 1951, Museum of Modern Art and Merchandise Mart, 1, MoMA online press archive, <https://www.moma.org/momaorg/shared/pdfs/docs/press_archives/1522/releases/MOMA_1951_0040.pdf> (accessed October 9, 2013).

12. "Good Design for 1949," *Interiors* 108 (December 1948), 114. Press release, June 22, 1951, Museum of Modern Art and Merchandise Mart, 1.

13. The tenth of "Twelve Precepts of Modern Design," in Edgar Kaufmann Jr., *What Is Modern Design?* (New York: Museum of Modern Art, 1950), 7.

14. Press release, June 22, 1951, Museum of Modern Art and Merchandise Mart.

15. Paul Reilly, "Presenting the Case for Furnishing: Vintage or Contemporary?," manuscript, 1958, 1–2, Paul Reilly, design administrator, papers, c.1920–c.1989, Archive of Art and Design, London.

16. Richard Hamilton, review of Arthur Drexler's *Introduction to Twentieth Century Design*, in *Design*, December 1959. Reyner Banham, "H.M. Fashion House," *New Statesman*, January 27, 1961, 151.

17. Daniel Horowitz, *Consuming Pleasures: Intellectuals and Popular Culture in the Postwar World* (Philadelphia: University of Pennsylvania Press, 2012).

18. Ibid., 2.

19. Ibid., 1.

20. Ibid., 2.

21. John E. Blake, "Consumers in Danger," *Design* 134 (February 1960), 25.

22. Ibid.

23. Ibid.

24. Michael Farr, *Design in British Industry: A Mid-Century Survey* (Cambridge: Cambridge University Press, 1955), xxxvi.

25. Patrick Maguire and Jonathan Woodham, *Design and Cultural Politics in Postwar Britain: The 'Britain Can Make It' Exhibition of 1946* (London: Leicester University Press, 1997).

26. Nikolaus Pevsner, "The Visual Arts," transcript of BBC Radio Third Programme, October 10, 1946, 4, BBC talks, 1946–1977, Series II, Box 52, Nikolaus Pevsner papers, 1919–1979, Getty Research Institute, Los Angeles.

27. John Ruskin, "Modern Manufacture and Design," Lecture to Bradford School of Design, 1859, in *The Works of John Ruskin*, vol. 10 (London: George Allen, 1878), 103.

28. William Morris, "Art and the Beauty of the Earth," lecture delivered at Burslem Town Hall, October 13, 1881, in William Morris, *Art and the Beauty of the Earth* (London: Longmans, 1898), 29.

29. Ibid., 23.

30. *Trend*, Spring 1936, 41–42.

31. Gillian Naylor, "Design Magazine, a Conversation, 22 September 2003," in *Design and the Modern Magazine*, ed. Jeremy Aynsley and Kate Forde (Manchester: Manchester University Press, 2007), 165.

32. Anthony Bertram, *The Enemies of Design* (Design and Industries Association, 1946), 3.

33. Obituary, Michael Farr, *Design*, October 1993, 6.

34. Interview with Michael Farr, 1991, Archive of Art & Design, London, AAD 7-1989, 42.

35. Farr, *Design in British Industry*, xxxvi.

36. D. M. Forrest, Commissioner, The Tea Bureau, "One More Word about Teapots," *Design* 82 (October 1955), 50.

37. "The Aims and Organization of the Council of Industrial Design," paper prepared for discussion at Regional Controllers' Office, on September 8, 1948, Paul Reilly, design administrator, papers, c.1920–c.1989, Archive of Art and Design, London.

38. Naylor, "Design Magazine, a Conversation," 165.

39. Council of Industrial Design Memorandum on A Design Centre for British Industry, ii, Paul Reilly, design administrator, papers, c.1920–c.1989, Archive of Art and Design, London.

40. Paul Burall, "The Official Critic: Irrelevant or Critical?," *Design Issues* 13, no. 2 (Summer 1997), 37.

41. "Interim Notes on the Editorial Contents of Design," prepared by Mass Observation Ltd., December 13, 1954, Design Council Archive, University of Brighton Design Archives, Brighton.

42. Ibid.

43. "A Report on Design Readership," prepared by Mass Observation Ltd., August 1961, Design Council Archive, University of Brighton Design Archives, Brighton.

44. John Blake, "The Case for Criticism," comment, *Design* 137 (May 1960), 43.

45. Ibid.

46. John Blake, "Towards a New Editorial Strategy," memo, October 1976, Design Council Archive, University of Brighton Design Archives, Brighton.

47. Farr, *Design in British Industry*, 208.

48. Stephen Spender, "Thoughts on Design in Everyday Life," design oration of the Society of Industrial Artists, 1958, published in *SIA Journal*, January 1959.

49. Ibid.

50. Richard Hamilton, personal interview, February 23, 2007.

51. David Robbins, ed., *The Independent Group: Postwar Britain and the Aesthetics of Plenty* (Cambridge, MA: MIT Press, 1990), 28.

52. Banham, "A Throw-away Esthetic," 64.

53. Lawrence Alloway, "The Long Front of Culture," *Cambridge Opinion*, no. 17 (1959), 25–26.

54. Richard Hamilton, personal interview, February 23, 2007.

55. Ibid.

56. The theme of the 1959 International Design Conference at Aspen was "Communication: The Image Speaks." The proceedings were published in the August 1959 issue of *Industrial Design*, and could, therefore, have informed Hamilton's lecture/article.

57. Richard Hamilton, "Persuading Image," *Design* 134 (February 1960), 32.

58. Ibid., 29.

59. Peter Drucker, "The Promise of the Next 20 Years," *Harper's Bazaar*, April 1955.

60. Peter Drucker quoted in Vance Packard, *The Waste Makers* (Brooklyn: Ig publishing, 2011), 66.

61. George Nelson, "Obsolescence," *Industrial Design*, December 1956, 88.

62. Ibid., 81.

63. Richard Hamilton, in "Persuading Image: A Symposium," *Design* 138 (June 1960), 57.

64. "Pop Art is: Popular (designed for a mass audience), transient (short term solution), expendable (easily forgotten), low cost, mass produced, young (aimed at youth), witty, sexy, gimmicky, glamorous, big business." Richard Hamilton, "Pop Art Is," unpublished, 1957, in Richard Hamilton, *Collected Words* (London: Thames & Hudson, 1983), 24.

65. Hamilton, in "Persuading Image: A Symposium," 57.

66. Richard Hamilton, "Artificial Obsolescence," in *Collected Words*, 155. Originally published in *Product Design Engineering*, January 1963.

67. Hamilton, "Persuading Image" (February 1960), 28–32.

68. Ibid.

69. Richard Hoggart, *The Uses of Literacy* (London: Penguin, 1992; first published London: Chatto & Windus, 1957).

70. "A Report on Design Readership," 13.

71. Hoggart, *The Uses of Literacy*, 215.

72. Richard Hamilton, "Popular Culture and Personal Responsibility," lecture at National Union of Teachers Conference, October 26–28, 1960, in Hamilton, *Collected Words*, 155.

73. Vance Packard, *The Hidden Persuaders* (London: Pelican, 1962), 16.

74. Hamilton, "Popular Culture and Personal Responsibility," 155.

75. Hamilton, "The Urbane Image," 53.

76. "Richard Hamilton in Conversation with Michael Craig-Martin, 1990," in *Richard Hamilton*, ed. Hal Foster with Alex Bacon (Cambridge, MA: MIT Press, 2010), 6.

77. John Blake, "Communication and Persuasion," *Design* 140 (August 1960), 34.

78. John Blake, "American Design in London," *Design* 200 (August 1965), 53.

79. Ibid.

80. Raymond Williams, *Britain in the Sixties: Communications* (London: Penguin, 1964), 75.

81. Daniel Boorstin, *The Image: A Guide to Pseudo Events in America* (New York: Vintage Books, 2012), 5.

82. Richard Hamilton, personal interview, February 23, 2007.

83. Ken Garland, quoted in Michael Farr, obituary, *Design*, October 1993, 6.

84. Ken Garland, personal interview, February 14, 2007.

85. Memos and Minutes of *Design* magazine meetings, 1954–1978, Design Council Archive, University of Brighton Design Archives, Brighton.

86. Memo from Mr. Tree to Mr. Hughes-Stanton, June 26, 1969, Memos and Minutes of *Design* magazine meetings, 1954–1978, Design Council Archive, University of Brighton Design Archives, Brighton.

87. Ken Garland, quoted in Michael Farr, obituary, *Design*, October 1993, 6.

88. Ken Garland, personal interview, February 14, 2007.

89. Interview with Michael Farr, August 15, 1991, Archive of Art & Design, AAD 7-1989. The collection of *Architectural Review* magazines housed in the V&A's National Art Library was donated by Michael Farr.

90. *Architectural Review* initiated a "Criticism" section in June 1951.

91. Memos and minutes of *Design* magazine meetings, 1954–1978, Design Council Archive, University of Brighton Design Archives, Brighton.

92. Ken Garland, personal interview, February 14, 2007. "Design Magazine, a Conversation," in Aynsley and Forde, *Design and the Modern Magazine*, 171.

93. Hamilton, "Persuading Image" (February 1960), 29.

94. Blake, "Consumers in Danger," 25.

95. John Blake, preface to "Persuading Image," *Design* 134 (February 1960), 28.

96. "Persuading Image: A Symposium," 54–57.

97. Ibid., 54.

98. Ibid., 55.

99. Ibid.

100. Ibid., 56.

101. Ibid., 55.

102. Ibid., 57.

103. John Hewitt, "Good Design in the Marketplace: The Rise of Habitat Man," *Oxford Art Journal* 10, no. 2 (1987), 30.

104. "A Report on Design Readership."

105. Ibid., 13.

106. Ibid., 14.

107. Ibid., 15.

108. Ibid.

109. In his article Hamilton made several references to the February 1959 issue of *Industrial Design* magazine. It was checked out of the Royal College of Art library repeatedly during March 1960, suggesting that his article created a new readership for the American magazine.

110. Richard Hamilton, personal interview, February 23, 2007.

111. Hamilton, "Popular Culture and Personal Responsibility," 154.

112. Ibid.

113. Reyner Banham, "Design by Choice," *Architectural Review*, July 1961, 44.

114. Ibid.

115. Jane Fiske Mitarachi, "Evaluating Industrial Design," *Journal of the American Association of University Women*, October 1958, 17.

116. Magazine subhead, *Industrial Design*, February 1954, 1.

117. Jane Thompson, personal interview, July 30, 2007.

118. Charles Whitney, "Publisher's Postscript," *Industrial Design*, February 1954, 150.

119. Subscription card, *Industrial Design*, February 1954.

120. George Nelson, "Planned Expansion," *Industrial Design*, February 1954, 148.

121. Arthur Drexler, quoted in "Design as Commentary," *Industrial Design*, February 1959, 56.

122. Ibid., 61.

123. William Snaith, quoted in "Design as Commentary," *Industrial Design*, February 1959, 60.

124. Their focus was primarily to introduce stricter codes of professional practice and to reinforce the legality of industrial design as a profession, established in a seminal case in 1940 where Teague successfully argued that it should be considered a profession in terms of taxation.

125. Henry Steele Commager, *The American Mind* (New Haven: Yale University Press, 1959), quoted in John Patrick Diggins, *The Promise of Pragmatism: Modernism and the Crisis of Knowledge and Authority* (Chicago: University of Chicago Press, 1994), 400.

126. Daniel Bell, *The End of Ideology: On the Exhaustion of Political Ideas in the Fifties* (New York: Free Press, 1960); Daniel J. Boorstin, *The Genius of American Politics* (Chicago: University of Chicago Press, 1953). For a detailed account of the influence of pragmatism in American culture, see Diggins, *The Promise of Pragmatism*.

127. Neil Jumonville, *Critical Crossings: The New York Intellectuals in Postwar America* (Berkeley: University of California Press, 1991), xii.

128. Jane Fiske McCullough, *Journal of the American Association of University Women*, October 1958, 14–15.

129. Ibid.

130. Jane Fiske McCullough, "Taste, Travel and Temptations," editorial preface, *Industrial Design*, July 1957, 25.

131. Jane Thompson, personal interview, July 30, 2007.

132. Jane Thompson, *Architecture Boston* 9, no. 4 (July/August 2006), 50.

133. Deborah Allen, personal interview, July 6, 2007.

134. Jane Thompson, personal interview, July 30, 2007.

135. Surprisingly, the British grant-funded magazine had better success with its advertising than the commercially driven US publication. Its advertisements were mainly for materials too, such as Pirelli rubber, Formica, and the British Aluminium Co., but they also managed to attract furniture companies like Hille, Knoll International, and Ercol, presumably because they were not in competition with an interior design magazine as *Industrial Design* was; but also, judging from the *Design* magazine memos and correspondence of the 1970s, *Design* had a comparatively aggressive sales staff.

136. Ralph Caplan, "I.D. Magazine, 1954–2009," *Voice*, AIGA website (January 5, 2010), < http://www.aiga.org/i-d-magazine-1954-2009/> (accessed September 17, 2012), para. 4.

137. Jane Thompson, *Architecture Boston* 9, no. 4 (July/August 2006), 50.

138. Jane Thompson, personal interview, July 30, 2007.

139. Jane Fiske Mitarachi and Deborah Allen, "The Trouble with Taxis," *Industrial Design*, February 1954, 11.

140. Deborah Allen, "The Body Beautiful: A Museum Asks 7 Men to Eye Automobiles," *Interiors*, May 1950, 112–116.

141. Walter Dorwin Teague, letter to the editors, *Industrial Design*, April 1958, 8.

142. Raymond Loewy, letter to the editors, *Industrial Design*, April 1954, 18.

143. Jane Thompson, personal interview, July 30, 2007.

144. Ibid.

145. Robert Osborn and Thomas B. Hess, "Who's Who in Distinguished Design," *Industrial Design*, February 1954, 68–71.

146. Jane Fiske McCullough, "Working in a Man's World," *Charm*, November 1957, 87.

147. Report on ASID's 14th annual conference, *Industrial Design*, June 1959, 60.

148. Jane Thompson, personal interview, July 30, 2007.

149. Ibid.

150. Deborah Allen, "The Driver's View: Cars '56," *Industrial Design*, August 1956, 138.

151. Jane Thompson, personal interview, July 30, 2007.

152. Jane Thompson, "Urbanist without Portfolio: Notes on a Career," in Claire Lorenz, *Women in Architecture USA* (New York: Rizzoli, 1990).

153. Recent work on the history of women's work has sought to dismantle the metaphor of the "female sphere," which had been used as a trope to characterize unequal power relations between the sexes, demonstrating instead the fluidity of interchange between the household and the world. See, for example, Linda K. Kerber, *Toward an Intellectual History of Women: Essays* (Chapel Hill: University of North Carolina Press, 1997).

154. Jane Thompson, personal interview, July 30, 2007.

155. Jane Fiske Mitarachi, "Critical Horseplay," editorial preface, *Industrial Design*, April 1957, 43.

156. Ibid.

157. Jane Fiske McCullough, review of *The Hidden Persuaders*, Books section, *Industrial Design*, May 1957, 10.

158. Jane Fiske Mitarachi, "Critical Horseplay," 43.

159. Jane Thompson, personal interview, July 30, 2007.

160. C. Edson Armi, *The Art of American Car Design: The Profession and Personalities* (University Park: Pennsylvania State University Press, 1989), 50.

161. Deborah Allen, "Crisis Year for Cars, Cars '58," *Industrial Design*, February 1958, 71.

162. Harrison E. Salisbury, *New York Times*, March 2, 1959, excerpted in *Industrial Design*, April 1959.

163. Raymond Loewy, *Never Leave Well Enough Alone* (New York: Simon & Schuster, 1951); Edgar Kaufmann Jr., "Borax or the Chromium Plated Calf," *Architectural Review*, August 1948, 88–92.

164. Edson Armi, *The Art of American Car Design*, 54.

165. Allen, "Cars '55," 82.

166. Deborah Allen, personal interview, July 6, 2007.

167. Deborah Allen, "The Dream Cars Come True Again," Design Review: Cars 1957, *Industrial Design*, February 1957, 103.

168. Ibid.

169. Allen, "Cars '55," 89.

170. Ibid.

171. Margaret Walsh, "Gender and Automobility: Selling Cars to American Women after the Second World War," *Journal of Macromarketing* 31, no. 1 (March 2011), 57–72.

172. Deborah Allen, personal interview, July 6, 2007.

173. Reyner Banham, "Vehicles of Desire," *Art*, September 1, 1955, 4.

174. Robert Hughes, "Only in America," *New York Review of Books*, December 20, 1979, <http://www.nybooks.com/articles/1979/12/20/only-in-america/> (accessed September 18, 2012), para. 18.

175. Allen, "Cars '55," 82.

176. Ibid.

177. Allen, "Crisis Year for Cars, Cars '58," 72.

178. Photography historian Tim Benton has observed how Banham, too, used the Wölfflinian technique of visual comparison: "For if Banham rejected parts of the high art history lecture, he was a master of the very Wölfflinian technique of visual comparison. We were all brought up in the tradition of the left and right projector screens and the basic grammar of art historical comparison. ... Selection and 'play' of images lies at the heart of this tradition and constitutes part of the argument." Tim Benton, "The Art of the Well-Tempered Lecture: Reyner Banham and Le Corbusier," in *The Banham Lectures: Essays on Designing the Future*, ed. Jeremy Aynsley and Harriet Atkinson, with a foreword by Mary Banham (Oxford: Berg, 2009), 11–32.

179. Rohm & Haas advertisement, *Industrial Design*, April 1957, 30–31.

180. Enjay Butyl advertisement, *Industrial Design*, April 1957, 29.

181. Allen, "Cars '55," 82.

182. Deborah Allen, "Cars '56: The Driver's View," *Industrial Design*, August 1956, 134.

183. Allen, "Crisis Year for Cars, Cars '58," 74.

184. Allen, "The Dream Cars Come True Again," 103.

185. Banham, "Vehicles of Desire," 3.

186. Ibid.

187. Reyner Banham, "The Atavism of the Short-Distance Mini-Cyclist," in *Design by Choice*, 88. Originally published in *Living Arts* 3 (1964), 91–97.

188. Reyner Banham, "Unlovable at Any Speed," *Architects' Journal* 144 (December 21, 1966), 1527–1529.

189. Reyner Banham, "Roadscape with Rusting Rails," *Listener* 80 (August 29, 1968), 267–268.

190. Banham, "Who Is This Pop?," 5.

191. Ibid., 13.

192. Banham, "Design by Choice," 44.

193. Banham, "Who Is This Pop?"

194. Vladamir Nabokov, *Lolita* (New York: Vintage Books, 1989), 141–142 (first published in Paris, 1955; in New York, 1958).

195. Reyner Banham, *Theory and Design in the First Machine Age* (London: Architectural Press, 1960), 37.

196. Mary Banham, personal interview, February 26, 2007.

197. Banham, "The Atavism of the Short-Distance Mini-Cyclist," 84. For a fuller account of Banham's cultural background and ideals in relation to class, see Nigel Whiteley, *Reyner Banham: Historian of the Immediate Future* (Cambridge, MA: MIT Press, 2002), 378–382.

198. Banham, "Who Is This Pop?," 13.

199. Banham, "Design by Choice," 48.

200. Banham, "A Throw-away Esthetic," 61.

201. Reyner Banham, "All That Glitters Is Not Stainless," *Architectural Design*, August 1967, 351.

202. Banham, "A Throw-away Esthetic," 64.

203. Leslie Fiedler, "The Middle against Both Ends," *Encounter*, August 1955, 16–23.

204. Ibid.

205. Gillian Naylor, "Theory and Design: The Banham Factor," *Journal of Design History* 10, no. 3 (1997), 245.

206. Banham, "Vehicles of Desire," 3.

207. Ibid.

208. Banham, "Design by Choice," 48.

209. Banham, "Vehicles of Desire," 4.

210. Banham, "Design by Choice," 43.

211. Banham, "A Throw-away Esthetic," 65.

212. Banham, margins, "Design by Choice," 48.

213. *Ulm 5*, quarterly bulletin of the Hochschule für Gestaltung, July 1959, 79.

214. Ibid.

215. Banham, "A Throw-away Esthetic," 65.

216. Reyner Banham, "Space for Decoration, A Rejoinder," *Design*, July 1955, 24.

217. Banham, "Vehicles of Desire," 4.

218. Ibid., 5.

219. Ibid.

220. Banham, "Who Is This Pop?," 3.

221. John J. Stinson, "Anthony Burgess: Novelist on the Margin," *Journal of Popular Culture*, Summer 1973, 136–151.

222. Anthony Burgess interviewed by John Cullinan, *Paris Review*, Spring 1973, <http://www.theparisreview.org/interviews/3994/the-art-of-fiction-no-48-anthony-burgess> (accessed September 20, 2012).

223. Banham's first piece for *New Society*, on August 19, 1965, would be "Kandy Kulture Kikerone," a review of Tom Wolfe's essay collection *The Kandy-Kolored Tangerine-Flake Streamline Baby* (New York: Farrar, Straus & Giroux, 1965). The essay after which the book was titled first appeared in *Esquire*, a magazine that Hamilton and Banham both read, in November 1963, as "There Goes (Varoom! Varoom!) That Kandy-Kolored (Thphhhhhh!) Tangerine-Flake Streamline Baby (Rahghhh!) Around the Bend (Brummmmmmmmmmmmmmmm) ..." and is regarded as the first product of the "New Journalism" genre.

224. Banham, "Vehicles of Desire," 3.

225. Banham, preface to *Design by Choice*, 7.

226. She decided to save his subsequent papers, and those written since 1976 are collected in the Getty Archive. Mary Banham, personal interview, February 26, 2007.

227. Ibid.

228. Ibid.

229. Banham, "A Throw-away Esthetic," 65.

230. Jane Fiske McCullough, "To the Reader," editorial preface, *Industrial Design*, March 1960, 35.

231. Don Wallance, letter to editors, *Industrial Design*, June 1960, 10. The Harvard economist John Galbraith critiqued the assumption that continually increasing material production is a sign of economic and societal health in his 1958 best-seller *The Affluent Society* (Boston:

Houghton Mifflin, 1958); and political scientist Walter Lippmann, who was awarded a Pulitzer Prize in 1958 for his syndicated column "Today and Tomorrow" in the *Herald Tribune*, was a prominent critic of the propagandist machinations of the mass media.

232. C. Wright Mills, "The Man in the Middle," *Industrial Design*, November 1958, 73.

233. Fred Eichenberger, letter to the editors, *Industrial Design*, June 1959, 8.

234. These thinkers included: the journalist Vance Packard, who leveled critiques at the advertising industry and its obsession with motivational research, which he held accountable for persuading people to buy things they didn't need (*The Hidden Persuaders*, 1957), and at American manufacturers for their adoption of planned obsolescence as a business model and consumers for their excessive consumption (*The Waste Makers: A Startling Revelation of Planned Wastefulness in Industry Today*, 1960); the politician Ralph Nader, whose investigations of deficiencies in American automobile design were published in 1959 and later republished in the book *Unsafe at Any Speed: The Designed-In Dangers of the American Automobile* (New York: Grossman, 1965); and the architect Richard Buckminster Fuller, whose writings about the wasteful practice of industrial design were collected in *No More Secondhand God and Other Writings* (Carbondale: Southern Illinois University Press, 1963).

235. Ralph Caplan, "On the Designer's Side," *Industrial Design*, February 1958, 33.

236. Banham, "Design by Choice," 44.

237. Ibid., 43.

CHAPTER 2

1. Ant Farm, biography, *Design Quarterly* 78/79 (1970), Special Double Issue on Conceptual Architecture, 10.

2. Steven Roberts, "The Better Earth: A Report on Ecology Action, a Brash, Activist, Radical Group Fighting for a Better Environment," *New York Times Magazine*, March 29, 1970, 8.

3. Chip Lord, personal interview, June 18, 2008.

4. Jean Baudrillard, *Utopia Deferred: Writings from Utopie 1967–1978*, trans. Stuart Kendall (New York: Semiotext(e), 2006), 77–78.

5. Michel de Certeau, "Making Do: Uses and Tactics," in *The Practice of Everyday Life* (Berkeley: University of California Press, 1984), 30.

6. *IDCA '70*, dir. Eli Noyes and Claudia Weill, IDCA, 1970.

7. Eli Noyes pursued a career in animation; Claudia Weill went on to direct documentaries and the 1978 hit movie *Girlfriends*.

8. Eli Noyes, personal interview, March 28, 2008.

9. Eli Noyes, personal correspondence, July 10, 2008.

10. Letter from Egbert Jacobson to Frank Stanton, President of Columbia Broadcasting System, January 2, 1951, International Design Conference in Aspen Records 1949–2006, 2007.M.7, Series 1, Box 1, Fol. 2, Getty Research Institute, Los Angeles.

11. Minutes of the planning meeting, February 19, 1951, International Design Conference in Aspen Records 1949–2006, 2007.M.7, Series 1, Box 1, Fol. 3, Getty Research Institute, Los Angeles.

12. Walter P. Paepcke, "The Importance of Design to American Industry," in promotional brochure for IDCA 1951, International Design Conference in Aspen papers, MSIDCA87, Box 15, Fol. 734, Special Collections and University Archives, University of Illinois at Chicago.

13. For a fuller account of the formation of the Aspen Institute for Humanistic Studies, see James Sloan Allen, *The Romance of Commerce and Culture: Capitalism, Modernism, and the Chicago-Aspen Crusade for Cultural Reform* (Boulder: University Press of Colorado, 2002).

14. Promotional brochure, IDCA 1951, International Design Conference in Aspen papers, MSIDCA87, Box 15, Fol. 734, Special Collections and University Archives, University of Illinois at Chicago.

15. Allen, *The Romance of Commerce and Culture*, 262.

16. Promotional brochure, IDCA 1957, International Design Conference in Aspen papers, MSIDCA87, Box 15, Fol. 736, Special Collections and University Archives, University of Illinois at Chicago.

17. "10th International Design Conference in Aspen," *Communication Arts*, July 1960.

18. "Tentative Program for the 1955 conference," memo, uncataloged, Aspen Institute.

19. Promotional brochure, IDCA 1961, International Design Conference in Aspen papers, MSIDCA87, Box 15, Fol. 740, Special Collections and University Archives, University of Illinois at Chicago.

20. Nikolaus Pevsner, "At Aspen in Colorado," *Listener*, 1953, republished in *The Aspen Papers*, ed. Reyner Banham (New York: Praeger, 1974), 16.

21. Reyner Banham, "A Private Memoir," in *The Aspen Papers*, 110.

22. *IDCA '70*, dir. Noyes and Weill.

23. "Aspen One-upmanship," Editorial in *Environment Planning and Design* (July/August 1970), 13.

24. Roberta Elzey, "Founding an Anti-University," in *Counter Culture: The Creation of an Alternative Society*, ed. Joseph Berke (London: Peter Owen, 1969), 244.

25. *Design Quarterly* 78/79 (1970), Special Double Issue on Conceptual Architecture, 6–10.

26. "Report on Long Range Planning of the IDCA," November 14, 1964, 2, uncataloged papers, Aspen Institute.

27. Philip B. Meggs, "Great Ideals: John Massey and the Corporate Design Elite," AIGA website, 1997,< https://www.aiga.org/medalist-johnmassey> (accessed November 10, 2016).

28. Minutes of the Board Meeting, June 1969, International Design Conference in Aspen papers, MSIDCA87, Box 2, Fol. 25, Special Collections and University Archives, University of Illinois at Chicago.

29. Ibid.

30. Minutes of Board Meeting, November 1969, International Design Conference in Aspen papers, MSIDCA87, Box 2, Fol. 25, Special Collections and University Archives, University of Illinois at Chicago.

31. Sim Van der Ryn, personal interview, June 18, 2008.

32. Ibid.

33. Minutes of Board Meeting, 1970, International Design Conference in Aspen papers, MSIDCA87, Box 2, Fol. 25, Special Collections and University Archives, University of Illinois at Chicago.

34. John Berger, "The Nature of Mass Demonstrations," in *Selected Essays*, ed. Geoff Dyer (New York: Vintage Books, 2001), 247.

35. Banham, "A Private Memoir," 111.

36. Mark Kurlansky, *1968: The Year that Rocked the World* (New York: Random House, 2004), 273.

37. Ibid.

38. Ibid., 274.

39. Memorandum to the IDCA Board of Directors, June 8, 1970, International Design Conference in Aspen papers, MSIDCA87, Box 3, Fol. 35, Special Collections and University Archives, University of Illinois at Chicago.

40. *IDCA '70*, dir. Noyes and Weill.

41. Ibid.

42. *1971 World Book Year Book* (Chicago: Field Enterprises Educational Corporation, 1971), 199.

43. Hubert Tonka, in *The Inflatable Moment: Pneumatics and Protest in '68*, ed. Marc Dessauce (New York: Princeton Architectural Press, 1999), 49.

44. Anty Pansera, "The Triennale of Milan: Past, Present, and Future," *Design Issues* 2, no. 1 (Spring 1985), 23.

45. "Milano 14 Triennale," *Domus* 466 (September 1968), 15.

46. Promotional brochure, IDCA 1962, International Design Conference in Aspen papers, MSIDCA87, Box 15, Fol. 741, Special Collections and University Archives, University of Illinois at Chicago.

47. Speaker biographies, IDCA 1970, International Design Conference in Aspen Records 1949–2006, 2007.M.7, Series 1, Box 28, Fol. 3, Getty Research Institute, Los Angeles.

48. Speakers' Papers, IDCA 1970, International Design Conference in Aspen Records 1949–2006, 2007.M.7, Series 1, Box 28, Fol. 4–6, Getty Research Institute, Los Angeles.

49. Ibid.

50. Audio cassette of IDCA 1970 proceedings, International Design Conference in Aspen papers, MSIDCA87, Box 11, Fol. 565, Special Collections and University Archives, University of Illinois at Chicago.

51. Speakers' Papers, IDCA 1970, International Design Conference in Aspen Records 1949–2006, 2007.M.7, Series 1, Box 28, Fol. 4–6, Getty Research Institute, Los Angeles.

52. *Student Handbook*, 1, IDCA 1970, International Design Conference in Aspen Records 1949–2006, 2007.M.7, Series 1, Box 28, Fol. 8, Getty Research Institute, Los Angeles.

53. William Houseman, "A Program Chairman's Diary of Sorts," Speakers' Papers, IDCA 1970, International Design Conference in Aspen Records 1949–2006, 2007.M.7, Series 1, Box 28, Fol. 4–6, Getty Research Institute, Los Angeles.

54. Ibid.

55. Les Levine, "Les Levine Comments on the IDCA," *Aspen Times*, June 25, 1970, 1-B.

56. Michael Doyle, *How to Make Meetings Work* (New York: Playboy Press, 1976); and *Meetings, Isn't There a Better Way?* (Visucom Productions, 1981).

57. Cliff Humphrey, "The Unanimous Declaration of Interdependence," in *Difficult but Possible: Supplement to the Whole Earth Catalog* (Menlo Park, CA: Portola Institute, September 1969), 12–13.

58. Ibid.

59. Samuel Ichiye Hayakawa, "How to Attend a Conference," Speakers' Papers, IDCA 1970, International Design Conference in Aspen Records 1949–2006, 2007.M.7, Series 1, Box 3, Fol. 6, Getty Research Institute, Los Angeles.

60. Ibid.

61. *IDCA '70*, dir. Noyes and Weill.

62. Reading List, Program, IDCA 1971, International Design Conference in Aspen Records 1949–2006, 2007.M.7, Series 1, Box 29, Fol. 8, Getty Research Institute, Los Angeles.

63. *IDCA '70*, dir. Noyes and Weill.

64. Ibid.

65. Letter from Reyner Banham to his wife, Mary, June 19, 1970, International Design Conference in Aspen Records 1949–2006, 2007.M.7, Series 1, Box 27, Fol. 3, Getty Research Institute, Los Angeles.

66. The French Group also included Eric Le Compte, an industrial designer at Eliot Noyes and Associates; Gilles de Bure, a design and media journalist who contributed to the CoID's *Design* magazine; and industrial designers Claude Braunstein and Roger Tallon and their wives, a professor of Greek and Latin literature and a physician, respectively. André Fischer, who read their statement at the closing session of the conference, is listed in the conference brochure as a "geographer." Each year from 1965 onward, an IBM International Fellowship had been awarded to a number of delegates from a foreign country to allow them to attend the conference. When Eliot Noyes asked the board to suggest a country for the 1970 conference, France had been proposed. There is no indication that France had been chosen because of the uprisings in Paris that put it at center stage of world politics in 1968. The logic had more to do with the fact that a country as influential as France, in terms of design and architecture, should no longer be overlooked.

67. Jean Baudrillard, "Statement Made by the French Group," Speakers' Papers, IDCA 1970, International Design Conference in Aspen Records 1949–2006, 2007.M.7, Series 1, Box 28, Fol. 4–6, Getty Research Institute, Los Angeles.

68. Ibid.

69. Ibid.

70. *IDCA '70*, dir. Noyes and Weill.

71. Felicity Scott, *Architecture or Techno-utopia: Politics after Modernism* (Cambridge, MA: MIT Press, 2007), 238.

72. Baudrillard, "Statement Made by the French Group."

73. Gilles de Bure, *C.R.É.É.*, no. 6 (November/December 1970), trans. Patricia Chen for *Rosa B* (2013).

74. Jean Baudrillard, interview with Jean-Louis Violeau, May 1997, in Baudrillard, *Utopia Deferred*, 18.

75. Baudrillard, "Statement Made by the French Group."

76. *IDCA '70*, dir. Noyes and Weill.

77. Letter from Reyner Banham to his wife, Mary, June 19, 1970.

78. "Resolutions by those attending the 1970 International Design Conference in Aspen, Friday, June 19, 1970, in recognition of our national—social—physical environment."

79. Ibid.

80. Stewart Udall, Speakers' Papers, IDCA 1970, International Design Conference in Aspen Records 1949–2006, 2007.M.7, Series 1, Box 28, Fol. 4–6, Getty Research Institute, Los Angeles.

81. Administrative and financial records, IDCA 1970, International Design Conference in Aspen Records 1949–2006, 2007.M.7, Series 1, Box 26, Fol. 1–5, Getty Research Institute, Los Angeles.

82. Letter from Reyner Banham to his wife, Mary, June 19, 1970.

83. Ibid.

84. Ibid.

85. Ibid.

86. Audio cassette, Summary, Michael Doyle, Fischer, Tabibian, Banham, 1970, International Design Conference in Aspen papers, MSIDCA87, Box 11, Fol. 578, Special Collections and University Archives, University of Illinois at Chicago.

87. Letter from Reyner Banham to his wife, Mary, June 19, 1970.

88. Ibid.

89. Tony Cohan, "Questions about Approach Plague Aspen," *Progressive Architecture*, August 1970, 39.

90. George Nelson, untitled lecture, Speakers' Papers, IDCA 1969, International Design Conference in Aspen Records 1949–2006, 2007.M.7, Series 1, Box 25, Fol. 1–5, Getty Research Institute, Los Angeles.

91. *Student Handbook*, IDCA 1970, International Design Conference in Aspen Records 1949–2006, 2007.M.7, Series 1, Box 28, Fol. 8, Getty Research Institute, Los Angeles.

92. Minutes of Meeting, IDCA 1970, p. 4, Minutes 1966–1973, International Design Conference in Aspen papers, MSIDCA87, Box 2, Fol. 25, Special Collections and University Archives, University of Illinois at Chicago.

93. Ibid.

94. Ibid.

95. Ibid.

96. *IDCA '70*, dir. Noyes and Weill.

97. Letter from Reyner Banham to his wife, Mary, June 19, 1970.

98. Banham, *The Aspen Papers*, 11.

99. Ibid., 207.

100. Reyner Banham, Manuscript, Aspen Reader, 1965–1973, Reyner Banham Papers, 1877–1988, undated, bulk 1960–1988, Series VI, Box 18, Fol. 1–4, Getty Research Institute, Los Angeles.

101. Banham, *The Aspen Papers*, 222.

102. Ibid., 223.

103. Dexter Masters, "Quick and Cheesy, Cheap and Dirty," in Banham, *The Aspen Papers*, 141.

104. Ibid.

105. Reyner Banham and Ralph Caplan, "The Aspen Papers," *Industrial Design*, August 1964, 58–61.

106. Ibid.

107. George Nelson, "The Rest of Our Lives," Speakers' Papers, IDCA 1969, International Design Conference in Aspen Records 1949–2006, 2007.M.7, Series 1, Box 25, Fol. 1–5, Getty Research Institute, Los Angeles.

108. Banham, *The Aspen Papers*, 92.

109. William Houseman, Speakers' Papers, IDCA 1969, International Design Conference in Aspen Records 1949–2006, 2007.M.7, Series 1, Box 28, Fol. 4–6, Getty Research Institute, Los Angeles.

110. Annotated Events List, Program, IDCA 1971, International Design Conference in Aspen Records 1949–2006, 2007.M.7, Series 1, Box 29, Fol. 8, Getty Research Institute, Los Angeles.

111. Richard Buckminster Fuller was not Farson's choice as a speaker. Saul Bass and Eliot Noyes had visited Farson during the planning of the conference to express their concern that the kinds of speakers he was enlisting were not recognized in the design community. They persuaded him to invite Fuller, who was liable to draw attendees. Farson recalls: "They thought it was going to go bust. I was used to being in a shapeless environment, but it was very difficult for them." Personal interview, June 30, 2008.

112. Promotional brochure, Publicity, IDCA 1971, International Design Conference in Aspen Records 1949–2006, 2007.M.7, Series 1, Box 29, Fol. 11, Getty Research Institute, Los Angeles.

113. Program, IDCA 1971, International Design Conference in Aspen Records 1949–2006, 2007.M.7, Series 1, Box 29, Fol. 8, Getty Research Institute, Los Angeles.

114. Ibid.

115. Richard Farson, personal interview, June 30, 2008.

116. Allan Kaprow, "Environments and Happenings" panel discussion, 1966, in Jeanne Siegel, *Artwords: Discourse on the 60s and 70s* (Ann Arbor, MI: UMI Research Press, 1985), 173, 169.

117. Richard Farson, personal interview, June 30, 2008.

118. Henri Lefebvre, *La Somme et le reste* (Paris: Éditions La Nef de Paris, 1959).

119. David Harvey, "Afterword," in Henri Lefebvre, *The Production of Space*, trans. Donald Nicholson-Smith (Oxford: Blackwell, 1991), 429.

120. Antoine Stinco in Dessauce, ed., *The Inflatable Moment*, 70.

121. Reading List, Program, IDCA 1971, International Design Conference in Aspen Records 1949–2006, 2007.M.7, Series 1, Box 29, Fol. 8, Getty Research Institute, Los Angeles.

122. Theodore Roszak, *The Making of a Counter Culture: Reflections on the Technocratic Society and Its Youthful Opposition* (London: Faber and Faber, 1972), xiii.

123. Richard Buckminster Fuller's geodesic domes, cartographic innovations, and visionary thinking inspired a generation of architects through such books as *Operating Manual for Spaceship Earth* (1969) and *Utopia or Oblivion: The Prospects for Humanity* (1972). Victor Papanek, faculty member and then Dean at CalArts and designer, wrote *Design for the Real World: Human Ecology and Social Change* (1972). This screed against unsafe and wastefully manufactured objects became a totemic title in the search for alternative design practices to suit an alternative lifestyle.

124. Maurice Stein, *Arts in Society* 7, no. 3 (Fall/Winter 1970), 64.

125. Registration brochure, Publicity, IDCA 1971, International Design Conference in Aspen Records 1949–2006, 2007.M.7, Series 1, Box 29, Fol. 11, Getty Research Institute, Los Angeles.

126. Richard Farson, personal interview, June 30, 2008.

127. Sheila Levrant de Bretteville, personal interview, May 14, 2008. "Some Aspects of Design from the Perspective of a Woman Designer," *Icographic: A Quarterly Review of International Visual Communication Design*, no. 6 (1973), 1–11.

128. Sheila Levrant de Bretteville, personal interview, May 14, 2008.

129. Richard Farson, personal interview, June 30, 2008.

130. Ken Margolies and Charlotte Gaines of the Parnassus Institute were asked to "direct a group of conferees in improvisational theater reflecting on the conference as a whole." "Annotated Events List," Program, IDCA 1971, International Design Conference in Aspen Records 1949–2006, 2007.M.7, Series 1, Box 28, Fol. 9, Getty Research Institute, Los Angeles.

131. Richard Farson, personal interview, June 30, 2008.

132. Farson and his wife still invited board members and other dignitaries to cocktails, at Trustee House #5, Aspen Meadows, for example.

133. Letter from Jack Roberts, IDCA president, to Richard Farson, January 18, 1971, Correspondence, International Design Conference in Aspen Records 1949–2006, 2007.M.7, Series 1, Box 29, Fol. 7, Getty Research Institute, Los Angeles.

134. Letters, Film Bookings, International Design Conference in Aspen papers, MSIDCA87, Box 1, Fol. 8, Special Collections and University Archives, University of Illinois at Chicago.

135. Jean Baudrillard, "Play and the Police," in *Utopia Deferred*, 36.

136. Minutes of the IDCA Board, June 22, 1974, Board of Directors, International Design Conference in Aspen papers, MSIDCA87, Box 2, Fol. 21, Special Collections and University Archives, University of Illinois at Chicago.

137. Matthew Holt, personal correspondence, May 10, 2012.

CHAPTER 3

1. Robin Young, "The Great Taste Test," *Times*, November 14, 1983, 7.

2. Hubert Dreyfus, Fernando Flores, and Charles Spinosa, *Disclosing New Worlds: Entrepreneurship, Democratic Action and the Cultivation of Solidarity* (Cambridge, MA: MIT Press, 1999), 19.

3. Bruno Latour and Peter Weibel, eds., *Making Things Public: Atmospheres of Democracy* (Cambridge, MA: MIT Press, 2005), 5.

4. Robin Kinross, "From Commercial Art to Plain Commercial," *Blueprint*, April 1988, 29.

5. *Blueprint*, for example, included a special supplement in its fifth anniversary issue in 1988 titled "The Design Decade," *Blueprint*, October 1988, 4.

6. Nigel Whiteley, *Design for Society* (London: Reaktion Books, 1993), 162.

7. John Butcher, Hansard, British Design Talent, HC Deb March 12, 1986, vol. 93 c928.

8. Editorial leader, *Blueprint*, October 1984, 3.

9. John Butcher, "Design and the National Interest: 1," in *Design Talks!*, ed. Peter Gorb (London: Ashgate, 1988), 218.

10. Editorial, *Blueprint*, October 1988, 4.

11. Robin Murray, "Life after Henry (Ford)," *Marxism Today*, October 1988, 11.

12. Kinross, "From Commercial Art to Plain Commercial," 29.

13. Paul Springer, email correspondence, July 16, 2012.

14. Celia Lury, *Consumer Culture* (New Brunswick, NJ: Rutgers University Press, 1996), 225.

15. Robert Elms, "1984: Style," *The Face*, January 1995, 51.

16. Peter York, "Style," in *Modern Times* (London: Futura, 1984), 9.

17. Ibid., 8.

18. Ibid., 9.

19. Ibid.

20. Meyer Shapiro, *Theory and Philosophy of Art: Style, Artist, and Society* (New York: George Braziller, 1994), 51. (Shapiro's essay on Style was first published in 1953 in *Anthropology Today*.)

21. Ibid.

22. Tom Wolfe, "The 'Me' Decade and the Third Great Awakening," *New York* magazine, August 23, 1976. Christopher Lasch, *The Culture of Narcissism: American Life in an Age of Diminishing Expectations* (New York: W. W. Norton, 1979).

23. *The Face*, December 1985, 46.

24. Pierre Bourdieu, *Distinction: A Social Critique of the Judgment of Taste*, trans. Richard Nice (Cambridge, MA: Harvard University Press, 1984), xi.

25. Ibid., 170–173.

26. Lury, *Consumer Culture*, 80.

27. Dick Hebdige, *Subculture: The Meaning of Style* (London: Methuen, 1979), 126.

28. Lury, *Consumer Culture*, 80.

29. Jane Thompson, lecture at SVA MFA Design Criticism, September 22, 2010.

30. Jane Thompson and Alexandra Lange, prologue to *Design Research* (San Francisco: Chronicle Books, 2010), 11.

31. Lury, *Consumer Culture*, 93.

32. "Designer Everything," *The Face*, September 1988, 154.

33. Frank Mort and Nicholas Green, "You've Never Had It So Good Again!," *Marxism Today*, May 1988, 32.

34. Judith Williamson, "The Politics of Consumption," in *Consuming Passions: The Dynamics of Popular Culture* (London: Marion Boyars, 1987), 233. First published in *New Socialist*, October 1985.

35. David Marquand, "The Enterprise Culture: Old Wine in New Bottles," in *The Values of the Enterprise Culture: The Moral Debate*, ed. P. Heelas and P. Morris (London: Routledge, 1992), 65.

36. Williamson, "The Politics of Consumption," 233.

37. Susan Sontag, *Illness as Metaphor and AIDS and Its Metaphors* (New York: Anchor Books, 1993), 63.

38. Ibid.

39. Sontag, *Illness as Metaphor and AIDS and Its Metaphors*, 153–154.

40. Martin Pawley, "This Is What Did Me In," *Blueprint*, June 1989, 10.

41. "Compulsive Shopping 'Real Illness,'" *Guardian*, October 6, 1994, quoted in Lury, *Consumer Culture*, 38.

42. Jon Wozencroft and Neville Brody, review, *Guardian*, December 2, 1988, 25.

43. Ibid.

44. Dick Hebdige, "World of Inferiors," *Blueprint*, May 1989, 40.

45. Williamson, *Consuming Passions*.

46. Peter York, "Anorexia of the Soul," in *Modern Times*, 62–67.

47. Dick Hebdige, *Hiding in the Light: On Images and Things* (London: Routledge, 1988), 209.

48. Nicola Roberts, "Black on Black," *New Musical Express*, July 19, 1986.

49. Deyan Sudjic, personal interview, June 1, 2010.

50. Ibid.

51. Peter Fuller, "Should Products Be Decorated?," *Design*, August 1983, 33.

52. Jane Lott, "Flip-top Torch," *Design*, May 1983, 49.

53. Jane Lott, "Walking Away from Violent Summer Shades," *Design*, May 1983, 13.

54. Peter Murray, email correspondence, May 2, 2016.

55. Deyan Sudjic, interview, in Liz Farrelly, "Design Journalism: The Production of Definitions," MA thesis, V&A/RCA, 1989, 39.

56. "Designers' Saturday," *Blueprint*, October 1983, 3.

57. Robin Kinross, introduction to *Unjustified Texts: Perspectives on Typography* (London: Hyphen Press, 2002), 13.

58. According to a 1987 profile, 67 percent of *Blueprint*'s readers were architects and interior designers, and 26 percent were "other designers."

59. Jennifer Scanlon, *Inarticulate Longings* (New York: Routledge, 1995).

60. Peter York, "Chic Graphique," in *Modern Times*, 36.

61. Simon Esterson, personal interview, August 5, 2010.

62. Deyan Sudjic, personal interview, June 1, 2010.

63. Fiona MacCarthy, letter, "The Bayley Spot," *Blueprint*, February 1984, 3.

64. Charles Jencks, quoted in Deyan Sudjic, "High Jencks," *Blueprint*, November 1984, 16.

65. Letter from Stephen Bayley, *Blueprint*, December/January 1984, 3.

66. *Blueprint*, June 1987, 3.

67. Ibid.

68. Katherine Hamnett, "The Politics of T-Shirts," *Blueprint*, October 1984, 11.

69. Peter York, personal interview, August 16, 2007.

70. Peter York, "The Meaning of Clothes," *Blueprint*, October 1983, 20.

71. Deyan Sudjic, "The Joy of Matt Black," *Blueprint*, November 1985, 42–45.

72. Deyan Sudjic, "Reinventing the Skyscraper," in *From Matt Black to Memphis and Back Again*, ed. Sudjic (London: Architecture Design and Technology Press, 1989), 12. Originally published in *Blueprint*, November 1985.

73. Ibid.

74. Richard Bryant, "Sex Tech," in Sudjic, ed., *From Matt Black to Memphis and Back Again*, 126. Originally published in *Blueprint*, March 1987.

75. Deyan Sudjic, "Inconspicuous Consumption," in Sudjic, ed., *From Matt Black to Memphis and Back Again*, 131. Originally published in *Blueprint*, July/August 1987.

76. Deyan Sudjic, "The Joy of Matt Black," *Blueprint*, November 1985, 44.

77. Ibid.

78. Ibid.

79. Deyan Sudjic, "How We Got from There to Here," *Blueprint*, September 1983, 16.

80. "Abolish the Design Council," editorial, *Blueprint*, December 1984/January 1985, 3.

81. Letters, *Blueprint*, February 1985, 9.

82. Ibid.

83. Sudjic, "High Jencks," 16.

84. Deyan Sudjic, *Cult Heroes: How to Be Famous for More Than Fifteen Minutes* (London: André Deutsch, 1989), 92. Excerpted in *Blueprint*, May 1989, 72.

85. Maurice Cooper, "The Deceptively Simple Style of Eva Jiricna," *Blueprint*, October 1983, 14.

86. Editorial, *Blueprint*, October 1983, 2.

87. Rick Poynor, "Grow Up *Blueprint*!," *Blueprint*, September 1993, 14.

88. Paul Atkinson and David Silverman, "Kundera's Immortality: The Interview Society and the Invention of the Self," *Qualitative Inquiry* 3, no. 3 (September 1997), 304–325.

89. Janet Abrams, "(Mis) Reading between the Lines," *Blueprint*, February 1985, 88.

90. Deyan Sudjic, "The Building Blocks of a Boy's Life," *Guardian*, April 30, 2006.

91. Deyan Sudjic, *Cult Objects: The Complete Guide to Having It All* (London: Paladin Granada, 1985).

92. Stephen Bayley, *The Good Design Guide: 100 Best Ever Products* (London: Conran Foundation, 1985).

93. Ibid., 3.

94. Ibid., 25.

95. Sudjic, *Cult Objects*.

96. Dick Hebdige, "World of Interiors," *Blueprint*, May 1989, 40.

97. Deyan Sudjic, interview, in Farrelly, "Design Journalism: The Production of Definitions," 28.

98. Letter from Tom Wolfe to Stephen Bayley, November 15, 1983, The V&A Archive, MA/28/387, Blythe House, London.

99. Robert Hewison, "Behind the Lines," *Times Literary Supplement*, March 25, 1983, 298.

100. Steve Wood, "The Man Who Walks Like a Trend," *Blueprint*, December/January 1983, 19.

101. Ibid.

102. Ibid.

103. Tom Wolfe, *The New Journalism* (London: Pan Books, 1990), 48, 36.

104. Ibid.

105. Ibid., 47.

106. Peter York, "Tom, Tom, the Farmer's Son," *Harpers & Queen*, October 1979, 210.

107. Ibid.

108. Ibid.

109. Deyan Sudjic, *Blueprint*, October 1988, 8.

110. Ibid.

111. Ibid.

112. Deyan Sudjic, "Milan: The Party Is Over," *Blueprint*, October 1983, 22.

113. Ibid.

114. Ibid.

115. Peter York, Introduction, in Sudjic, ed., *From Matt Black to Memphis and Back Again*, 6. The ABC-TV series *Thirty Something* (1987–1991) features the advertising mogul Miles Drentell (played by David Clennon). This character "wears expensive suits, strokes a Zen sandbox, and speaks in a terrifyingly snide, controlled monotone," and the interior of the office is a Hollywood composite of details gathered from research into the design and advertising industry. William Drenttel, a designer with experience in advertising, provided much of the information, in conversation with his college friend Edward Zwick, the series producer. "We were white and generally male. We bought our (white) shirts at one of three places: Brooks Brothers, J. Press ('of New Haven'), or Paul Stuart. There were no other acceptable choices." William Drenttel, "I Was a Madman," *Design Observer*, July 11, 2008, <http://designobserver.com/feature/i-was-a-mad-man/6997/> (accessed October 15, 2016).

116. Deyan Sudjic, "Is There Life After Sloane?," *Blueprint*, November 1984, 11.

117. Peter York, personal interview, August 16, 2007.

118. Ibid.

119. Simon Esterson, personal interview, August 5, 2010.

120. Ibid.

121. *Skyline* was published by the Institute for Architecture and Urban Studies in New York, 1979–1983.

122. The culture of filing magazines in design studios was well established. *Architectural Review* provided their issues with holes pre-punched in sections that would be added to subject-specific ring binders in studio libraries.

123. Deyan Sudjic, interview, in Farrelly, "Design Journalism: The Production of Definitions," 45.

124. "The Design Decade," *Blueprint*, October 1988, 4.

125. Ibid.

126. Editorial, *Blueprint*, December 1989, 7.

127. Sudjic, *Cult Heroes*.

128. Ibid., 10.

129. Deyan Sudjic, "The Design Decade," *Blueprint*, October 1988, 4.

130. Editorial, *Blueprint*, September 1988, 7.

131. Sudjic, *Cult Heroes*, 109.

132. Deyan Sudjic, interview, in Farrelly, "Design Journalism: The Production of Definitions," 45.

133. Deyan Sudjic, Editorial, *Blueprint*, September 1988, 7.

134. Brian O'Docherty, *Inside the White Cube: The Ideology of the Gallery Space* (Berkeley: University of California Press, 1999).

135. Stephen Bayley, *In Good Shape: Style in Industrial Products 1900–1960* (London: Design Council Books, 1979).

136. James Woudhuysen, "Acquired Taste," *Blueprint*, October 1983, 17.

137. Stephen Bayley, personal interview, January 6, 2011.

138. O'Doherty, *Inside the White Cube.*

139. *New Musical Express*, July 19, 1986.

140. Peter York, "Chic Graphique," in *Modern Times*, 26.

141. Brian Sewell, *The Tatler*, September 1983.

142. Jules Lubbock, "Style Victim," *The New Statesman*, June 7, 1985, 38.

143. Jean Baudrillard, "The Ecstasy of Communication," in *The Anti-Aesthetic: Essays on Postmodern Culture*, ed. Hal Foster (New York: New Press, 1998), 145–154.

144. Ibid.

145. Stephen Bayley, personal interview, January 6, 2011.

146. Marina Vaizey, "V&A Tastes," *Sunday Times*, October 2, 1983.

147. Barty Phillips, *Conran and the Habitat Story* (London: Weidenfeld and Nicolson, 1984), 103.

148. Stephen Bayley, personal interview, January 6, 2011.

149. See Christopher Frayling, *Henry Cole and the Chamber of Horrors: The Curious Origins of the V&A* (London: V&A Publishing, 2010).

150. Robert Hewison, *Culture and Consensus: England, Art and Politics since 1940* (London: Methuen, 1995), 271–272.

151. "A Tale of Two Cities" advertisement for Butler's Wharf, *Blueprint*, October 1989, 42–43.

152. Stephen Bayley, personal interview, January 6, 2011.

153. "He Sells Living in Style," *Reader's Digest*, February 1982, 51.

154. Stephen Bayley, personal interview, January 6, 2011.

155. Judith Williamson, personal interview, August 4, 2010.

156. Stephen Bayley, *Taste: An Exhibition about Values in Design* (London: Conran Foundation, 1983), 11.

157. This incident was reported in numerous publications, including *Blueprint*: "Farrell's Fury," "Sour Grapes," *Blueprint*, October 1983, 26.

158. Colin Amery, *Financial Times*, September 19, 1983.

159. Anne Engel, "Young Bayley and The Boilerhouse," *Harpers & Queen,* February 1981, 45. The Keep Britain Tidy campaign, initiated by the Women's Institute in 1955, became a limited company in 1984.

160. Gert Selle, "There Is No Kitsch, There Is Only Design," *Design Issues* 1, no. 1 (1984), 49.

161. Stephen Bayley, personal interview, January 6, 2011.

162. Mary Douglas, *Purity and Danger: An Analysis of the Concepts of Pollution and Taboo* (New York: Routledge, 2000), 2–6.

163. Ibid.

164. Young, "The Great Taste Test," 7.

165. Roland Barthes, "Sapanoids and Detergents," in *Mythologies* (New York: Hill and Wang, 2012), 33. Stephen Bayley was reported to have discussed this article over lunch with Robert P. Wilkinson, director of external affairs for Coca-Cola, at Alastair Little's restaurant in Soho. The details of their lunch (fillet of brill with mustard seeds, Perrier, Californian Sauvignon) and their topics of conversation were recounted in Amanda Craig, "Waiter, There's a Fly on the Wall," *Tatler*, May 1986, 120.

166. Judith Williamson, "Three Kinds of Dirt," in *Consuming Passions*, 223–227.

167. Ibid., 225.

168. Ibid., 227.

169. Mary Blume, *Herald Tribune*, November 18, 1983, 7.

170. Stephen Bayley, *Sunday Express Magazine,* September 18, 1983, 23.

171. Fiona MacCarthy, "A Lot of Things," *London*, June 1989, 8.

172. Stephen Bayley, personal interview, January 6, 2011.

173. Gavin Stamp, "Hard Boiled and Half Baked," *Spectator*, February 27, 1982.

174. Roy Strong, quoted in Young, "The Great Taste Test," 7.

175. Peter York, script, "Kaleidoscope," BBC Radio 4, December 14, 1983, 7, Design Museum archive, uncataloged.

176. Young, "The Great Taste Test," 7.

177. Patrick Wright, *A Journey through the Ruins: The Last Days of London* (Oxford: Oxford University Press, 2009), 37.

178. Stephen Bayley, quoted in Deyan Sudjic, "Stephen Bayley's Guide to Best-Selling Taste," *Sunday Times Magazine*, September 11, 1983.

179. Barbara Usherwood, "The Design Museum: Form Follows Funding," *Design Issues* 7, no. 2 (Spring 1991), 87.

180. Stephen Bayley, "A Haven for Modern Muses," in *Stephen Bayley: General Knowledge* (London: Booth Clibborn Editions, 2000), 210. Originally published in *Weekend Guardian*, July 1–2, 1989.

181. Wright, *A Journey through the Ruins*, 37–38.

182. Deyan Sudjic, "Commercial but Cultured," *Blueprint*, September 1989, 62.

183. Ibid.

184. Dick Hebdige, personal interview, April 3, 2011.

185. Ibid.

186. Ibid.

187. Dick Hebdige, "Some Sons and Their Fathers," in *The Impossible Self* (Winnipeg: Winnipeg Art Gallery, 1988), 71–82.

188. Ibid., 72.

189. Dick Hebdige, personal interview, April 3, 2011.

190. Hebdige, "A Report on the Western Front," *Block* 12 (1986/7), 4–26.

191. Fredric Jameson, "Postmodernism and Consumer Society," in Foster, ed., *The Anti-Aesthetic*, 127–144. Originally delivered as a lecture at the Whitney Museum, autumn 1982.

192. Dick Hebdige, "The Impossible Object: Towards a Sociology of the Sublime," *New Formations*, no. 1 (Spring 1987), 69.

193. Baudrillard, "The Ecstasy of Communication," 132.

194. Hebdige, "A Report on the Western Front," 6.

195. Ibid., 6–7.

196. Peter Barry, *Beginning Theory: An Introduction to Literary and Cultural Theory* (Manchester: Manchester University Press, 2009), 61.

197. Hebdige, "A Report on the Western Front," 7.

198. Ibid., 11.

199. Ibid., 12.

200. Ibid., 13.

201. Ibid.

202. Ibid., 14.

203. The Habitat catalog was sold for £1.25 in the early 1980s, double the price of other lifestyle magazines such as *The Face*. James Woudhuysen, "A Matter of Taste," *Designing*, No. 1, Design Council, 1983, 104.

204. Hebdige, "A Report on the Western Front," 19.

205. Ibid.

206. Ibid., 20.

207. Ibid.

208. Ibid.

209. Ibid., 21.

210. Dick Hebdige, personal interview, April 3, 2011.

211. Dick Hebdige, "Shopping for Souvenirs in the Occupied Zone," *Blueprint*, December 1988/January 1989, 12.

212. Dick Hebdige, "Shopping-Spree in Conran Hell," *Block* 15, 1989, 61.

213. Ibid., 56.

214. Ibid., 57.

215. Ibid., 60.

216. Cooper, "The Deceptively Simple Style of Eva Jiricna," 15.

217. Hebdige, "A Report on the Western Front," 19.

218. Judith Williamson, introduction to *Consuming Passions*, 13.

219. Williamson, "The Politics of Consumption," 230.

220. Williamson, introduction to *Consuming Passions*, 13.

221. Judith Williamson, in an interview with Gerry Beegan, *dot dot dot*, no. 4. Dot dot dot website (accessed October 9, 2013).

222. Judith Williamson, foreword to *Deadline at Dawn: Film Criticism 1980–1990* (London: Marion Boyars, 1993), 9.

223. Judith Williamson, personal interview, August 4, 2010.

224. Editorial, *City Limits*, October 1981.

225. Judith Williamson, personal interview, August 4, 2010.

226. Williamson, foreword to *Deadline at Dawn*, 10.

227. Judith Williamson, personal interview, August 4, 2010.

228. Judith Williamson, "Viewfinder," in *Deadline at Dawn*, 21. Originally published in *New Statesman*, December 12, 1986.

229. Ibid., 22.

230. Ibid., 15.

231. Ibid., 5.

232. Judith Williamson, "Belonging to Us," in *Consuming Passions*, 206. Originally published in *City Limits*, 1983.

233. Judith Williamson, introduction to *Deadline at Dawn*, 15.

234. Williamson, foreword to *Deadline at Dawn*, 8.

235. Judith Williamson, personal interview, August 4, 2010.

236. Ibid.

237. Williamson, "The Politics of Consumption," 231.

238. Judith Williamson, personal interview, August 4, 2010.

239. Williamson, "Three Kinds of Dirt," 226.

240. Judith Williamson, "Urban Spaceman," in *Consuming Passions*, 211.

241. Ibid.

242. Ibid., 201.

243. Hewison, *Culture and Consensus*, 212.

244. Sudjic, *Cult Objects*, 37.

245. Bayley, *The Good Design Guide*, 83.

246. "Walkman Receives a Belting Innovation," *Planner*, March 31, 1982.

247. Jean Baudrillard, *America* (London: Verso, 2010), 39. Originally published 1986.

248. Judith Williamson, personal interview, August 4, 2010.

249. Williamson, "Urban Spaceman," 210.

250. Ibid.

251. Williamson, "The Politics of Consumption," 232.

252. Ibid.

253. Judith Williamson, personal interview, August 4, 2010.

254. Williamson, "Belonging to Us," 205.

255. Ibid., 206.

256. Judith Williamson, personal interview, August 4, 2010.

257. Ibid.

258. Williamson, "The Politics of Consumption," 229.

259. Judith Williamson, personal interview, August 4, 2010.

260. Judith Williamson, "Textuality: On Postmodern Reading," Books section, *Marxism Today*, September 1991.

261. Ibid.

262. Judith Williamson, personal interview, August 4, 2010.

263. Judith Williamson, "Lost in the Hypermarket," *City Limits*, December 1988, 12.

264. Ibid., 13.

265. Introduction, special issue on "New Times," *Marxism Today*, October 1988, 3.

266. Judith Williamson, "Even New Times Change," *New Statesman & Society*, July 7, 1989, 33.

267. Ibid., 35.

268. Dick Hebdige, "After the Masses," *Marxism Today*, January 1989, 56.

269. Sudjic, "How We Got from There to Here," 17.

270. Barbara Radice, *Memphis: Research, Experiences, Results, Failures and Successes of New Design* (New York: Rizzoli, 1984), 187.

271. Sudjic, "The Design Decade," 4.

272. Stephen Bayley, *Commerce and Culture* (London: Design Museum, 1989), 83.

273. Ibid.

274. Stephen Bayley, personal interview, January 6, 2011.

275. Jon Wozencroft and Neville Brody, review, *Guardian*, December 2, 1988, 25.

276. Peter Murray, personal interview, November 10, 2008.

CHAPTER 4

1. "Shopkeeper extraordinaire Terence Conran sees good design as a social issue and wants 'everyone to have a better salad bowl.'" Regina Nadelson, "The Emperor of the Everyday," *Metropolitan Home*, February 1983, 47.

2. Anthony Dunne, *Hertzian Tales: Electronic Products, Aesthetic Experience and Critical Design* (London: RCA CRD Research Publications, 1999).

3. Paul Virilio, *The Information Bomb* (London: Verso, 2005), 9.

4. Andrew Rawnsley, *Servants of the People: The Inside Story of New Labour* (London: Penguin, 2001); and *The End of the Party: The Rise and Fall of New Labour* (London: Penguin, 2010).

5. Mark Leonard, *Britain™: Renewing our Identity* (London: Demos, 1997), 6.

6. Ibid., 3.

7. Ibid.

8. The report was based on work Leonard had previously done with the Design Council—a discussion paper titled "Views on Britain's Identity" (Design Council, 1997).

9. Michael Brenson, "The Curator's Moment," *Art Journal* 57, no. 4 (Winter 1998).

10. Paul O'Neill, "The Curatorial Turn: From Practice to Discourse," in *Issues in Curating Contemporary Art and Performance*, ed. Judith Rugg and Michele Sedgwick (Bristol: Intellect Books, 2008), 13.

11. Lord Clinton Davis to the House of Lords, March 30, 1998, HL Deb 30 March 1998, vol. 588cc11–2WA, <http://hansard.millbanksystems.com/written_answers/1998mar/30/powerhouseuk-exhibition> (accessed October 7, 2013).

12. Panel 2000, a Foreign and Commonwealth Office initiative, launched on the same day as "powerhouse::uk," was intended to produce "a strategy to improve the way Britain is seen overseas." Its panelists included John Sorrell, director of the Design Council and commissioner of the Demos report, as well as Mark Leonard, the report's author, along with industrialists and MPs.

13. Giles Worsley, "Portrait of the Artist as an Architect," *Daily Telegraph*, July 17, 1999.

14. Nigel Coates, quoted in "Cool Britannia Hits the Street," BBC News website, April 3, 1998.

15. John Battle, quoted in "Cool Britannia Hits the Street."

16. Sophie Barker, "Rebranding: PR Caution over Cool as a Corporate Tool," *PR Week*, April 24, 1998.

17. Nonie Niesewand, "Britain's Export Showcase Is Hot Air," *Independent*, March 27, 1998; Philip Browning, "Blairite Britain Enshrined in a Bouncy Castle," *New York Times*, April 15, 1998.

18. Jonathan Glancey, "Repacking Britain," *Guardian*, April 2, 1998, 4.

19. Judith Williamson, "Inflated Intangibles," *Eye*, Spring 1999, 7.

20. Ibid.

21. Glancey, "Repacking Britain," 4.

22. Claire Catterall, personal interview, September 17, 2007.

23. In notes for the "Stealing Beauty" exhibition, under the question "What do Perrier-Jouët want out of this?," one of the answers was: "to be seen to have integrity in their understanding of design." "Stealing Beauty" files, uncataloged, ICA Archive.

24. "Stealing Beauty" files, uncataloged, ICA Archive. "Commerce and Culture" exhibition budget, uncataloged, Design Museum archive. "powerhouse::uk" budget, uncataloged, personal papers.

25. "Stealing Beauty" press update, "Stealing Beauty" files, uncataloged, ICA Archive.

26. Claire Catterall, *Stealing Beauty: British Design Now* (London: ICA, 1999), 7.

27. Rita Felski, *Doing Time: Feminist Theory and Postmodern Culture* (New York: New York University Press, 2000), 79.

28. See in particular Henri Lefebvre, *Critique of Everyday Life*, vol. 1, trans. John Moore (London: Verso, 1991).

29. Notes in "Stealing Beauty" files, uncataloged, ICA Archive.

30. Ibid.

31. Felski, *Doing Time*, 81.

32. Penny Sparke, quoted in David Redhead, "The Irresistible Rise of the Anonymous," *Blueprint*, September 1993, 79.

33. Rick Poynor, "When Too Much Is Too Much," *Blueprint*, October 1997, 36–37.

34. Ibid.

35. Ibid.

36. John Thackara, preface to *Eternally Yours: Visions on Product Endurance*, ed. Ed van Hinte (Rotterdam: 010 Publishers, 1997).

37. Paul Neale, interview with Putri Trisulo, March 2, 2011, in "Curating Now," MA thesis, RCA/V&A, 2001.

38. Redhead, "The Irresistible Rise of the Anonymous," 77.

39. Ibid.

40. Claire Catterall, personal interview, September 17, 2007.

41. Nicolas Bourriaud, *Relational Aesthetics*, trans. Simon Pleasance and Fronza Woods (Dijon: Les presses du réel, 2002), 110. Originally published in French in 1998.

42. Tord Boontje, "Rough and Ready," sheet of explanatory notes, "Stealing Beauty" files, uncataloged, ICA Archive.

43. Claire Bishop, *Participation* (London: Whitechapel, 2006), 10.

44. Catterall, *Stealing Beauty*, 8.

45. Claire Catterall, "New Graphic Realism," *Blueprint*, November 1998, 34.

46. "Steal Yourself," *Wallpaper*, April 1999, 154.

47. Richard Benson, "Folk," *The Face*, January 2000, 83.

48. FAT, in answer to the question "What is your worst design memory?," in Catterall, *Stealing Beauty*, 25.

49. Wolfgang Tillmans had a solo show in 1997 at the Chisenhale Gallery in London. During the spring of 1999, the *Guardian*'s "Designer Living" column featured several of the abodes of the "Stealing Beauty" designers.

50. Nick Currie, "A Duchamp Moment," *Frieze*, August 6, 2008, <https://frieze.com/article/duchamp-moment>, accessed October 19, 2016.

51. Giles Reid, "Gruel Britannia," *Object*, no. 6 (1999), 29.

52. Catterall, *Stealing Beauty*, 9.

53. Paul O'Neill, *The Culture of Curation and the Curation of Culture(s)* (Cambridge, MA: MIT Press, 2012), 26.

54. Gareth Williams, "Design in a Dilemma," *Blueprint*, May 1999, 71.

55. Letters from ICA exhibition organizer Katya Garcia-Anton to the exhibited designers, "Stealing Beauty" files, uncataloged, ICA Archive.

56. Claire Catterall, personal interview, September 17, 2007.

57. Ibid.

58. Claire Catterall, working notes for "Stealing Beauty," then titled "All Kinds of Everything Remind Me of You," fax, July 29, 1998, "Stealing Beauty" files, uncataloged, ICA archive.

59. "Stealing Beauty" (Things to know in my absence) by David Wilkingson, "Stealing Beauty" files, uncataloged, ICA Archive.

60. Claire Catterall, personal interview, September 17, 2007.

61. Hal Foster, "Hey, That's Me," *London Review of Books*, April 5, 2001, 13.

62. Charles Leadbeater, "The New Entrepreneurism," *Guardian*, February 7, 2000. See also Charles Leadbeater, *Living on Thin Air: The New Economy* (London: Penguin, 2000).

63. Bruce Ferguson, "Exhibition Rhetorics," in *Thinking about Exhibitions*, ed. Reesa Greenberg, Bruce Ferguson, and Sandy Nairne (London: Routledge, 1996), 178.

64. Catterall, *Stealing Beauty*, 9.

65. Claire Catterall, quoted in Vanessa Pawsey, "Just a Feeling," *Design Week*, April 9, 1999, 36.

66. Fiona Raby, personal interview, July 21, 2011.

67. Anthony Dunne and Fiona Raby, "Hertzian Tales and Other Proposals," in *Frequencies: Investigations into Culture, History and Technology*, ed. Melanie Keen (London: Iniva, 1998), 46.

68. Anthony Dunne and Fiona Raby, quoted in "Product Overload: Designers Fight Back," *Blueprint*, October 1997, 38.

69. Dunne, preface to *Hertzian Tales*.

70. Alex Seago and Anthony Dunne, "New Methodologies in Art and Design Research: The Object as Discourse," *Royal College of Art Research Papers* 2, no. 1 (1996/1997), 3.

71. Anthony Dunne, personal interview, July 21, 2011.

72. Anthony Dunne, "Design Noir," *Blueprint*, November 1998, 25.

73. Dunne, *Hertzian Tales*, 51.

74. Anthony Dunne, "Form Follows the Software," review of "Metropolis: Tokyo Design Visions," *Blueprint*, December–January 1992, 44.

75. Anthony Dunne, quoted in "Product Overload," *Blueprint*, October 1997, 38.

76. Dunne, "Design Noir," 25.

77. Fiona Raby, personal interview, July 21, 2011.

78. Dunne, *Hertzian Tales*, 92.

79. Seago and Dunne, "New Methodologies in Art and Design Research," 1.

80. Ibid., 4.

81. Ibid.

82. Dunne & Raby credit two exhibitions in particular, held at the RCA in 1996—"This Appliance Might be Earthed" and "Monitor as Material"—for helping them to gain visibility for their work, and attract the attention of the design media and curators from the V&A and the British Council.

83. Fiona Raby, personal interview, July 21, 2011.

84. Ibid.

85. Dunne, *Hertzian Tales*, 58.

86. Anthony Dunne and Fiona Raby, *Design Noir: The Secret Life of Electronic Objects* (London: August, 2001), 75.

87. Anthony Dunne, personal interview, July 21, 2011.

CHAPTER 5

1. NA Cat Lover, "An Expensive Way to Smell Poo," Amazon, 2007, <http://www.amazon.com/review/R32RN0APNFZAUM> (accessed April 23, 2016).

2. Ibid.

3. Geoff Dyer, *But Beautiful* (New York: Picador, 1996), vii.

4. B. Govern, "Dual Function Design," Amazon, 2008, <http://www.amazon.com/review/R2XKMDXZHQ26YX/ref=cm_cr_rdp_perm?ie=UTF8&ASIN=B000OE2OLU> (accessed April 23, 2016).

5. "How Digg Works," Digg.com, 2007, <https://web.archive.org/web/20070228155957/http://digg.com/how> (accessed April 23, 2016).

6. Edgar, "Make This Your Only Stock and Store," Amazon, 2008, <http://www.amazon.com/review/RXXPVOUH9NLL3> (accessed April 23, 2016).

7. B. Smith, "Forum," Amazon.com, 2008, <http://www.amazon.com/forum/-/Tx2N-RE61HPIN09P/ref=ask_ql_ql_al_hza?asin=B00032G1S0> (accessed April 23, 2016).

8. Mike Mozart, "Dora the Explorer AquaPet FAIL TOYS Funny Naughty Dora Toy Review by Mike Mozart of JeepersMedia," YouTube.com, 2008, <https://www.youtube.com/watch?v=EgBvkHwCxJc> (accessed April 22, 2016).

9. Ralph Caplan, introduction to Judith Ransom Miller, "The History of Boys' Socks, 1947–1957," *Industrial Design*, June 1958, 54.

10. Reyner Banham, "O Bright Star," *New Society* 63, no. 1068 (May 5, 1983), 188–189.

11. Jane Thompson, personal interview, July 30, 2007.

12. Rebecca Blood, Weblogs: a history and perspective, September 7, 2000, <http://www.rebeccablood.net/essays/weblog_history.html> (accessed November 20, 2016).

13. Sabine Himmelsbach, "Blogs: The New Public Forum," in *Making Things Public: Atmospheres of Democracy*, ed. Bruno Latour and Peter Weibel (Cambridge, MA: MIT Press, 2005), 919.

14. Maggie Jackson, *Distracted: The Erosion of Attention and the Coming Dark Age* (Amherst, MA: Prometheus Books, 2008), 13.

15. Andrew Keen, *The Cult of the Amateur: How Today's Internet Is Killing Our Culture and Assaulting Our Economy* (London: Nicholas Brealey, 2007), 2.

16. Ibid., 38.

17. Jodi Dean, *Blog Theory: Feedback and the Capture of the Circuits of Drive* (Cambridge, UK: Polity Press, 2010), 40.

18. Ibid., 33.

19. Ibid., 83.

20. Armin Vit, "Bio," Speak Up, 2003, <http://web.archive.org/web/20030110104655/http://www.underconsideration.com/speakup/> (accessed October 31, 2013).

21. Mark Kingsley, "Rick Poynor: Ipse Dixit," Speak Up, May 7, 2007, <http://web.archive.org/web/20070531222459/http://www.underconsideration.com/speakup/archives/003354.html#138338> (accessed October 31, 2013).

22. Rick Poynor, "Easy Writer," *Print* 61, no. 3 (May/June 2007), 33–34.

23. Ibid.

24. Ibid.

25. Ibid.

26. Ibid.

27. Rick Poynor, "The Death of the Critic," Icon website, March 2006, <http://www.iconeye.com/read-previous-issues/icon-033-|-march-2006/the-death-of-the-critic-|-icon-033-|-march-2006> (accessed October 31, 2013).

28. Mark Kingsley, personal interview, November 13, 2012.

29. Ibid.

30. Mark Kingsley, "Rick Poynor: Ipse Dixit," Speak Up, May 7, 2007, <http://web.archive.org/web/20070531222459/http://www.underconsideration.com/speakup/archives/003354.html#138338> (accessed October 31, 2013).

31. Ibid.

32. Ibid.

33. Mark Kingsley, personal interview, November 13, 2012.

34. Ibid.

35. Mark Kingsley, comment, Speak Up, May 4, 2007, in response to Mark Kingsley, "Rick Poynor:Ipse Dixit," Speak Up, May 4, 2007, <http://web.archive.org/web/20070531222459/http://www.underconsideration.com/speakup/archives/003354.html#138338> (accessed October 31, 2013).

36. Armin Vit, comment, Speak Up, May 6, 2007, in response to Mark Kingsley, "Rick Poynor: Ipse Dixit," Speak Up, May 4, 2007, <http://web.archive.org/web/20070531222459/http://www.underconsideration.com/speakup/archives/003354.html#138338> (accessed October 31, 2013).

37. Dean, *Blog Theory*, 48.

38. ps, comment, Speak Up, May 11, 2007, in response to Mark Kingsley, "Rick Poynor: Ipse Dixit," Speak Up, May 4, 2007, <http://web.archive.org/web/20070531222459/http://www.underconsideration.com/speakup/archives/003354.html#138338> (accessed October 31, 2013).

39. Rick Poynor, "Poynor Replies to Speak Up's Discussion of 'Easy Writer,'" *Print* website, May 11, 2007, <http://web.archive.org/web/20070515072818/http://www.printmag.com/design_ articles/poynor_easy_writer/tabid/221/Default.aspx#response> (accessed October 31, 2013).

40. Marian Bantjes, comment, Speak Up, May 5, 2007, in response to Mark Kingsley, "Rick Poynor: Ipse Dixit," Speak Up, May 4, 2007, <http://web.archive.org/web/20070531222459/ http://www.underconsideration.com/speakup/archives/003354.html#138338> (accessed October 31, 2013).

41. Rick Poynor, personal interview, July 13, 2007.

42. Joe Natoli, comment, Speak Up, May 18, 2007, in response to Mark Kingsley, "Rick Poynor: Ipse Dixit," Speak Up, May 4, 2007, <http://web.archive.org/web/20070531222459/http://www. underconsideration.com/speakup/archives/003354.html#138338> (accessed October 31, 2013).

43. Hans Magnus Enzensberger, "Constituents of a Theory of the Media," in *Critical Essays* (New York: Continuum, 1982), 73. This essay was originally published in 1970.

44. Armin Vit, "Speak Up: Now What?," Speak Up, February 13, 2007, <http://web.archive. org/web/20070224215334/http://www.underconsideration.com/speakup/archives/003011. html> (accessed October 31, 2013).

45. Armin Vit, "Goodbye, SpeakUp," Speak Up, April 13, 2009, <http://www.underconsideration .com/speakup/archives/006034.html> (accessed October 31, 2013).

46. Rick Poynor, "Adventures in the Image World," Design Observer, November 5, 2010, <http://observatory.designobserver.com/entry.html?entry=21758> (accessed October 31, 2013).

47. Dean, *Blog Theory*, 40.

48. Ibid.

49. Thomas Friedman, "Dancing in a Hurricane," *New York Times*, November 19, 2016, 1.

50. Joe Clark, "The Death of Graphic Design Criticism," category on his personal website, <http://blog.fawny.org/category/graphic-design/dcrit/> (accessed October 31, 2013). Thomas Fisher, "The Death and Life of Great Architecture Criticism," Places, Design Observer, December 1, 2011, https://placesjournal.org/article/the-death-and-life-of-great-architecture-criticism/ (accessed October 19, 2016). Rick Poynor, "Another Design Voice Falls Silent," Design Observer, December 11, 2011, <http://observatory.designobserver.com/feature/another-design-voice-falls-silent/31828/> (accessed October 31, 2013). Rick Poynor, "The Death of the Critic," Icon website, March 2006, <http://www.iconeye.com/read-previous-issues/icon-033-|-march-2006/the-death-of-the-critic-|-icon-033-|-march-2006> (accessed October 31, 2013). Rick Poynor, "Where Are the Design Critics?," Design Observer, September 25, 2005, <http:// observatory.designobserver.com/entry.html?entry=3767> (accessed October 31, 2013).

51. 2008 saw the launch of an MFA in Design Criticism at the School of Visual Arts in New York and an MA in Design Writing Criticism at the University of the Arts in London.

52. Boris Groys, *Art Power* (Cambridge, MA: MIT Press, 2008), 111.

53. Michael Rock, in Michael Rock and Rick Poynor, "What Is This Thing Called Graphic Design Criticism?," *Eye* 4, no. 16 (Spring 1995), 57.

54. Clay Shirky, *Here Comes Everybody* (London: Penguin Books, 2008), 55.

55. Nicholas Negroponte, *Being Digital* (New York: Alfred Knopf, 1995), 153.

56. Brian Thill, *Waste* (New York: Bloomsbury Academic, 2015), 25.

INDEX

"Designing happens via critique, but for too long design criticism has been a distinct, weak discourse. *Sifting the Trash* articulates what robust historically situated design criticism can and must now be."
Cameron Tonkinwise, Professor of Design, University of New South Wales Art and Design, Sydney, Australia

"With vivid prose and fresh, compelling illustrations, *Sifting the Trash* presents a perceptive history of late twentieth-century British and American design criticism. Alice Twemlow uses a case study approach to trace shifts in critical emphasis from moralizing about design, to warning the public about its insidious influence, to promoting an open DIY approach."
Jeffrey L. Meikle, Stiles Professor in American Studies, University of Texas at Austin

"Alice Twemlow's stratigraphy of design criticism from its emergence in the 1950s to the present brilliantly exposes and explores the nascent discipline's struggle to balance the demands for product promotion with those for social critique. If *Sifting the Trash* is a history of the emergence of the design critic qua professional figure, it is no less a nuanced assessment of the role's fragility in an era in which consumers have been recast as curators and critics of a vastly expanded world of products."
Jeffrey T. Schnapp, Faculty Director, metaLAB, Harvard University